THE BIG BAY HORSE DOESN'T LIVE HERE ANYMORE

THE BIG BAY HORSE DOESN'T LIVE HERE ANYMORE

A Memoir

By Nell Parsons

©2019 by Nell Parsons

All rights reserved. This book or any portion thereof may not be reproduced or used in any manner whatsoever without the express written permission of the publisher except for the use of brief quotations in a book review.

ISBN:978-1-08131-716-4

FOREWORD

Nell was fifty-nine years old when she became the proud owner of a tall, handsome show horse. Marlon was seven years old and a winner in the show ring, but he was a flawed, high-strung, mad-at-the world horse whose many struggles had left him starved for affection.

It was a long, terrifying journey for Nell to earn his trust and to convince him she was on his side. Her earlier life challenges – being the first in her family to go to college, becoming a military wife amidst family squabbles and watching her newborn struggle for life in a military hospital – gave her the courage she needed to never give up on Marlon.

From growing up in rural Alabama in the 1940's to taking on horse showing with her husband in her senior years, Nell invites readers into her story of commitment, love, loss and adventure.

CONTENTS

Chapter 1	My Adventure Begins, 1991	1
Chapter 2	Growing Up on a Farm	9
Chapter 3	Searching for My Horse, 1993	19
Chapter 4	Adjusting to Life-Altering Events	33
Chapter 5	How Marlon Changed My Life	51
Chapter 6	Becoming Parents, 1957	55
Chapter 7	Facing Another Life-Altering Decision	67
Chapter 8	Riding Lesson	83
Chapter 9	First Horse Clinic	87
Chapter 10	Mississippi Horse Show	95
Chapter 11	Entering the Beginner Classes	103
Chapter 12	Riding to Avoid Hitting the Judge	113
Chapter 13	Marlon's Behavior, Always Surprising	119
Chapter 14	Church, the Cowboy Way	137
Chapter 15	Marlon Bolts, and I Am Thrown	141
Chapter 16	My Granddaughters	145
Chapter 17	Fiftieth Wedding Anniversary	155
Chapter 18	Going to Horse Shows	165
Chapter 19	Training Maggie	175
Chapter 20	Marlon Falls to His Knees	181
Chapter 21	My Perfect Ride	185
Chapter 22	Unusual Things Scare Marlon	189

Chapter 23	My Last Wild Ride	195
Chapter 24	Bob's Surgery and the Horse Commercial, 2007	201
Chapter 25	Jeannene's Visit	207
Chapter 26	Trip to Germany and Bob's Illness, 2009	217
Chapter 27	First Year Alone	227
Chapter 28	Diane and Julianne Move in with Me, 2010	237
Chapter 29	Goodbye, Marlon and Dan, 2010	243
Chapter 30	Going Forward: Trip to China, 2011	253
Chapter 31	John, 2013	259
About the Author		275

PROLOGUE

When you drive up to where I live, you will see pastures lush, green, and well kept. They are enclosed by a white fence matching the white trim on the yellow house. A nondescript but very functional barn sits inside the fence.

The pastures are freshly mowed, and the grass along the miles of fencing has been recently trimmed. You feel there isn't a sprig of grass growing where it shouldn't be. In plain sight is a yellow pump house with white trim near one of the fences. A long row of pine trees borders each side, providing plenty of shade.

Gates are everywhere, closing off or giving access from one area to the next. Let your eyes drift farther, and you will see two more pastures with big shade trees in each. A wood fence with a gate divides them. One section of this fence is missing. It was shattered into pieces by the big bay horse who doesn't live here anymore. After months of investigating, no one could figure out why he did it.

There is a large, red, round pen filled with truckloads of sand to provide good footing. The sand stayed loose and porous with the many circles the big bay horse made while walking, trotting, and cantering inside this ring. In the pastures, hoofprints reveal big circles, little circles, straight lines, casual walks along the fence lines, and fast gallops down the middle, all made by the big bay horse that doesn't live here anymore.

Now the barn is closed, empty, and still. The only sign of life is the skin shed by the huge black snake still living on the property. Not a creature can be seen hanging out under the pine trees or having a cool drink of water drawn from the well inside the yellow pump house with the white trim.

If you took a jaunt through the pastures, you wouldn't see any hoofprints or chunks of earth dug up by the big bay horse when he'd decide to leap into the air from a standstill and do his acrobatic moves. His specialties were sailing through the air, doing body twists and turns with his head and tail held high, and adding kicks in every direction. Yet he still managed to look regal with every movement while adding squeals of delight to entertain you. Time stood still then. You laughed out loud with pure joy, and your troubles melted away.

But all is silent now, as the big bay horse doesn't live here anymore.

CHAPTER 1
MY ADVENTURE BEGINS, 1991

"I want to get a winning quarter horse, learn to ride, and compete in horse shows," my husband announced one day. This was an astounding statement, even laughable, considering he was fifty-nine years old, couldn't ride, and didn't seem inclined to do the work involved in such a big undertaking. After recovering from the shock, I mostly ignored what he had said. In the days and weeks that followed, Bob kept repeating this statement. He eventually became angry when he wasn't taken seriously.

"Do you know where you can get such a horse?" I finally asked.

"A quarter horse farm is located on Sand Mountain, near your old home place. It's about an hour's drive from here," he replied.

Forty or so years ago, that farm was in the middle of nowhere. I agreed to go with him to look for the farm near Henagar, Alabama. I still had reservations about this new journey Bob wanted to launch, knowing that if one person has a horse, everyone in the family is involved. This totally occupied my thoughts as we drove through the rural countryside to locate the farm.

We arrived and discovered a working quarter horse farm with horses of all ages and two nationally known horse trainers with a slew of riders working for them. The large barn had two long aisles lined with stalls on both sides; at one end, there was an area for washing horses. A necessary tack room used for storing riding equipment was located just inside the front entrance, and next to it were a lounge and a bathroom with showers.

One of the workers took us outside to see extra stalls for the overflow of horses. Pastures nearby held mothers with babies; another pasture held the older babies that were being weaned. Outdoor riding rings and a big indoor arena for riding young or unruly horses and for tending them when the weather was bad completed the busy scene.

We were accustomed to seeing large barns for boarding horses but not a big training operation like this. Bob and I kept whispering to each other, oohing and aahing about what we were seeing. Besides all the horses and riders, we saw workers busy keeping the place running. Stalls were being cleaned, horses rotated in and out to graze, and hay put in stalls for later. Our escort told us the fences, arenas, paddocks, and pastures must be checked and maintained daily. There was a barn manager to oversee this operation. Dogs and cats of all varieties lived at the barn. They happily welcomed all newcomers. After greeting us, they followed us around until they heard another car arrive with new visitors.

If this wasn't dazzling enough, customers were milling around looking and checking out horses. Some were parents watching their children have riding lessons, while others were just bystanders out on a lovely fall day for a little amusement. We watched and chatted with a few parents and bystanders. We discussed prices with one father trying to get a ballpark price of a show horse. "The price you pay is nothing compared to the long-term upkeep of a horse," he said. This was information to think about.

The trainer indicated that they had a few horses suitable for Bob. Besides being different ages, they came in a variety of colors. The majority were sorrel, bay, gray, black, or palomino. There was only one stallion but plenty of mares and lots of geldings, which are castrated stallions. Bob didn't get on a horse that day. We just enjoyed looking at the magnificent horses and watching the trainers ride some of them.

A whole new world opened up for us. It lit a fire under Bob that couldn't be extinguished without him owning a horse. We made an appointment to come back so that Bob could check out a few of the horses for sale. My thoughts and emotions swung back and forth like a pendulum. I loved the atmosphere at the farm, the excitement of the clients, and the thrill of seeing such graceful horses at work. But I worried about Bob getting one at his age and not being able to ride.

Returning to the farm a week later, Bob was shown three young sorrel geldings. I was terrified these horses were too young and didn't have enough training for someone just beginning. I voiced my opinion; no one listened. The geldings were saddled and brought out to the arena. The trainers rode them around, showed us how they moved, and demonstrated what they could do. They explained that the horses should move forward in a steady, fluid motion at the walk, trot, and lope. A rider gives the horse cues by applying pressure with their legs and heels on the sides of the horse. It takes practice for horse and rider to understand each other.

On this sunny afternoon, as we watched these gorgeous creatures working in the outdoor arena, Bob was thrilled, even though he was nervous about the adventure. I watched and wavered between awe, wonder, and sheer terror for him. "These horses are too young for you," I kept repeating. It is a common saying that horses don't have any sense until they are about seven years old. By this age, they have experienced more of the world and should have received more schooling.

By my insistence, Iris, a five-year-old mare, was brought out, and one of the trainers rode her. She was older than the geldings and appeared to be calm and steady.

"Do you want to ride?" the trainer asked Bob after she finished riding.

Bob was scared and worried, but outwardly he appeared calm. He'd worn his western boots and jeans and was dressed for the challenge. He climbed into the saddle. The stirrups were adjusted, and the trainers gave suggestions on how he should cue the horse. Bob and Iris started walking around the arena. Spectators in the vicinity suddenly noticed a new rider was trying out a horse. They came closer to the open riding area, hanging back at a safe distance, pretending they weren't really interested in what was going on. The mare began to pick up speed without being asked. Trainers yelled directions to Bob—do this, do that! Iris got faster and faster. By now, they were totally out of control. Another trainer yelled for Bob to pull back on the reins. When he did, one rein broke, throwing him off balance.

Iris started bucking. Bob hit the ground. He landed flat on his back, taking the brunt of the fall on his tailbone.

My terrible prediction had happened, and he'd been on the horse I had chosen. This wasn't good! All sorts of horrible thoughts ran through my head. One was a picture of the actor Christopher Reeve falling from his horse. I was worried Bob might be paralyzed or have some other serious injury. Everyone rushed over to see how he was. He stood up and could walk, but there was pain in his lower back.

As we were talking to him and assessing the damage, a FedEx driver, one of the nonwatchers nearby, ambled over. "I give you a total of eighty points," he said. "You get fifty points for staying on eight seconds; you only get thirty points for how well you rode."

He was referring to how bull riders are scored. They are given fifty points for staying on eight seconds and can earn up to an additional fifty points depending on how well they ride. Everyone laughed. Bob was hurting, and he said he didn't know whether to give him a punch or to laugh. He did manage a chuckle.

There would be no more riding today. When we got home and looked at Bob's lower back, we were shocked to see a huge, swollen bump about the size of a saucer. I convinced him to go see a doctor.

"My, that is impressive!" the doctor said when she saw his injury. After an X-ray, she reported, "Nothing is broken. You will have to take some medicines for comfort while waiting for it to heal."

It took a long time for the swelling to go down, and he carried a sizable bump permanently. We found out later that Iris didn't like male riders. To my amazement, Bob never mentioned that I had insisted he try this older horse.

When Bob felt better, we went back to the farm for him to check out the three geldings that I'd been afraid for him to try. The trainers brought them out, again demonstrated how each moved, and showed how far along they were in their training. They all seemed about equal in looks and movement as far as we could tell with our limited ability to judge a good horse. Cathy, one of the trainers, told Bob which gelding was the best of the three, and he was encouraged to try it out.

If Bob was reluctant to get on a young horse, he didn't show it. He climbed into the saddle, and the horse, Dan, calmly walked around the ring. After several times around, Bob asked him to trot. He wanted to get past the trot after his previous ride. Dan behaved perfectly. The tension left Bob's face, and a smile replaced it.

His mind was made up. He bought Dan, a registered quarter horse whose official name was Royal's Dandy Step. The quarter horse is normally a heavily muscled, compact horse known for running the quarter mile faster than other breeds. It is associated with rodeos and other western events. They are known for their calm, laid-back personality. Bob learned to ride and never fell off Dan, proving I didn't know anything when it came to horses. The day Bob bought him, he left Dan with the trainer for additional training. He drove up several times a week after work for riding lessons. He went on weekends if the trainers were there. I often went with him to observe. The lessons were hard work and scary at times, but the challenge and excitement were enough for Bob to keep working at it. This became the dominant topic of conversation at our house.

Bob discovered that the trainers went to horse shows most weekends. There wasn't anyone to give him riding lessons on his off days from work. As soon as he had a little riding experience, he was encouraged to take Dan to the shows. In the beginning, he went to expose Dan to the atmosphere and activities on the showgrounds.

Always in a hurry, Bob quickly moved into the novice western pleasure class, ready or not. Requirements were to dress in western clothes, groom Dan to perfection, treat the saddle and bridle leathers with oils, and polish the silver. Bridles and western saddles were loaded with silver for decoration. I inherited the job of doing most of these chores and became the number-one horse holder. Bob was the worst in wanting someone to hold his horse while he was waiting for his novice class to begin. Riders had to be ready and waiting in the holding area. There wasn't an exact time for a class to begin; it depended on how long the previous classes had lasted. Bob didn't want to be tied down holding his horse. He wanted to wander around, talk to people about riding techniques, compare trainers, and speculate on what the day's judges were looking for.

As my horse-holding duties lasted longer and longer, I became unhappy. But my complaining fell on deaf ears. Bob was having too much fun.

One day at a horse show, I strolled into a tack shop, and the proprietor started talking to me. "What do you do?" she asked. "Do you have a horse?" This finalized my idea that it was time to get my own horse. You are nobody at a horse show if you don't have a horse. When I announced to Bob and Cathy, the professional horse trainer, that I wanted my own horse, they readily agreed with me.

My decision was made. It was time to start looking for my horse. Two years had passed since Bob had purchased Dan. Now I was in the same situation as he was then, fifty-nine years old and couldn't ride. I wanted a calm, older horse with years of training.

CHAPTER 2
GROWING UP ON A FARM

I grew up on a farm where we had working mules, and our survival depended on them. My family couldn't have taken better care of Littlen and Old Nell if they'd been made of gold, but I never considered them playthings. After working with the mules six days a week, we were happy to have a day of rest from each other. If someone had suggested riding one of them for fun, I would have thought them insane.

After going away to college at seventeen and never living on the farm again, my view of the world changed. A few years after leaving home, I met and married Bob. Months after our wedding, he entered the army, and we spent the next twenty-nine years moving around the United States and other countries. I soon discovered that my new husband was a dedicated animal lover. When our son arrived, the pressure was on to get him a puppy. A daughter came along three years later, and Bob started saying, "Every little girl needs a kitten of her very own." Year by year, animals kept coming.

We moved to Colorado Springs when our son, Rob, was twelve years old, and Diane, our daughter, was nine. We arrived at the beginning of summer without knowing anyone. For a summer activity, I signed the children up to go spend one day a week at a horse ranch. The ranch was a scene straight from the *Little House on the Prairie* television show. Windswept, barren land on the side of a sloping hill went as far as the eye could see. Large and small corrals and a dilapidated barn were located beyond a small house. Several trees surrounded a swimming pool, and barnyard dogs announced our arrival. Mrs. Gee, a weathered little lady, greeted us. She explained that the children rode horses in the mornings and swam in the hot afternoons. A van would transport them to and from the ranch.

Rob and Diane came home exhausted from their weekly trip to the ranch. Diane never quit talking about Supressa, the old horse that was selected for her to ride. She was a safe horse for a new rider, as she never took one loping step. She was past that stage of her life. Diane fell in love with horses. Rob enjoyed riding as well as the other activities.

"Any little girl who is that crazy about horses deserves to have one once in her life," Bob was saying a few months later.

Owning horses and moving every year or two didn't go together, but as he presented his case, I began to think of my favorite saying: "You only live once!" This gave me permission to reevaluate my stand on any subject.

The search began, and soon we added a very old horse to our menagerie of pets. Diablo was a good beginner horse.

When we brought Diablo to the ranch, Mrs. Gee and all the boarders were there to welcome him and see how he would handle joining the twenty-one other horses in the pasture. On the ranch, there were only a few ramshackle stalls for the old and ailing horses, plus a large shed for storing hay. All the other horses went into one large pasture. They had to find their place in the horse pecking order.

Mrs. Gee had two five-year-old geldings fighting for dominance over the herd and causing constant turmoil. She was hoping a new, older horse would establish order, prevent injuries, and calm all the horses.

Feeding time arrived, and Mrs. Gee loaded her rusty old truck with enough hay to make twenty-two separate piles, one for each horse. She drove into the pasture, and the gate was closed behind her. As she followed a straight line down through the pasture, a worker unloaded big piles of hay every few feet. Diablo stood and observed the geldings jockeying for the first pile. Then he lowered his head with a mean look on his face, and with confidence, he marched up and started eating. The geldings backed away. It was his attitude that convinced them that he was serious and would fight, if necessary. The geldings continued to battle among themselves for second place, keeping the other horses confused. Diablo kept his head turned so he could observe the herd. If he felt one of them was getting close and might challenge him, he'd quickly turn around and throw a kick in their direction with his huge, pie-plate-size feet. The younger horses weren't happy. It took them a long time to accept that today they couldn't win. There was much scuffling around and threatening each other as each horse tried to find his place in the new pecking order. On and on down the line it went until all the horses had a pile of hay. As the first horse finished his pile, he moved to the next one and made that horse shift farther down the line. This continued until every blade of hay was gone, or the first horses were full and moved away. The lowest horse in the pecking order must eat fast because he was the last to get his hay and the first horse to be permanently run off.

Diablo ate his hay and watched. This scenario played out at every feeding until they stopped challenging him for first place. Soon this wise old horse had established order among the whole herd.

The previous owners told us Diablo was twelve years old. But when the vet came to give him shots, he estimated Diablo was at least twenty-five years old. Hearing this, I nearly had a heart attack. I felt he was too old to be ridden. The vet assured me he was in good shape. Diablo proved to be a master at looking after himself. If the children wanted him to run more than he was willing, he'd pick up the lope, then stop suddenly and lower his head. The children flew over his head and hit the ground. Asking him to jog around too much would cause him to run under low tree limbs and knock them off. When they went to the pasture to catch him, he wouldn't cooperate that day if he didn't feel up to playing with them. It was always best to heed his guidance. On one of these days, Rob went to catch Diablo. As the horse was walking away, Rob grabbed his mane and hung on. Diablo moved faster and faster. Rob finally let go, rolling several times. The bridle with the iron bit, which he'd been holding in his other hand, hit him in the head, cutting a deep gash. He had to go to the emergency room for stitches. This smart ranch horse taught his first-time horse owners, as well as a pasture full of horses, so many things that we needed to know.

Soon after we bought Diablo, Diane and I went to the ranch to get better acquainted with him. He was on top of the hill at the far end of the pasture with the other twenty-one horses. Since he didn't know us, I decided to take a feed bucket with grain.

As we trudged up the long, steep hill, the horses spotted us and shifted their grazing positions so they could keep us in sight. Halfway there, a few horses started moving in our direction. I asked Diane to move a safe distance away from me. Jockeying began among the horses, and in a matter of seconds, they were running down the hill. Picking up speed, they galloped straight toward me. I was petrified and thought, "What can I do? I can't outrun them." Desperate, I held the feed bucket far out to my right just before they reached me. The lead horse veered to my right, and the herd followed. They had missed me, but as turmoil ensued, they regrouped and prepared for another run for the bucket. Watching them, I turned around and held the bucket far to my left. This time they were smarter and didn't run as fast or go as far. Shifting the bucket to the right, to the left, even up over my head, I quickly realized it was a losing battle. By now, horses were milling all around me, grabbing for the bucket. Terrified and shaking in my boots, I tossed the bucket as far as possible and ran for my life.

After I reassured Diane that all was well, we started walking back down the hill without the bucket or Diablo. I kept looking back at the horses scuffling over their surprise afternoon snack. Although shaken up over the incident, deep down I knew I should have known better. It was a much longer, disappointing walk back than our happy, hopeful walk up the hill to see our new horse.

Arriving back at the barnyard, we found Mrs. Gee waiting for us. She had magically appeared, and there was no doubt that she had been watching the whole time. Seeing my dejected demeanor, she offered words of encouragement. I was angry that she hadn't warned me against walking into a herd of horses with a bucket of feed. In later years, I discovered horse owners let new owners learn most horse things on their own. There are two reasons for this: a new horse owner will never forget what they learn by experience, and it is such fun for the old timers to watch, laugh, and reminisce. I suspected she had come out to console me so she wouldn't lose her new boarder.

The children took turns going to the ranch each day. It wasn't long before Bob decided we needed to add a younger horse so the kids could swap days riding the old horse and the young one. Both were better riders by now.

It didn't take long to find the perfect horse. At least this was what I was told. Pokey, a five-year-old registered quarter horse, joined us. He was solid black with four white stockings. Years later, because of his coloring, Fort Myer Army Base wanted to buy him to pull their funeral caissons. We felt insulted. How could they want our Pokey for such a sad job?

Soon orders for Alaska arrived, and we had only a couple of months to prepare for the move. After much anxiety and trying to figure out our options, we decided to buy a pickup truck and put a small camper on the back. This would make it easier to take the dog and cat with us. Diablo was too old to travel such a long distance, and the ranch mistress wanted to keep him to control her pasture of horses.

Our next problem was what to do with the younger horse we had recently bought. Many tears were shed over what to do about Pokey. We couldn't leave him behind; he didn't have a home to go to.

"Besides having a job, I am busy preparing to move the family to Alaska, trading the van for a heavy-duty truck that will survive the Alaskan highway, and searching for a camper to fit on the back of the truck. If you can find a way to get Pokey to Alaska, I will be happy to send him," Bob finally said to me.

A new challenge arose. I started contacting tack shops and stables for information about hauling horses, and I discovered that the Air Force Academy had stables. I called, and the manager told me a lady who had recently brought her horse down the Alaskan highway was there at the moment. He let me speak to her. She gave me the name and telephone number of a man in Anchorage, Alaska, who was coming to Salt Lake City in a couple of weeks to pick up a load of horses.

I called him, and he said, "I have room for one more horse."

A week before we left for the 4,500-mile drive to Alaska, we borrowed a trailer. For days Bob had been practicing driving our new truck with the camper on the back. With a rented hitch, we added the horse trailer, but there wasn't much time left for him to drive it around. At the end of the week, we drove Pokey to Salt Lake City. It was a long day, as we made a couple of wrong turns. Bob didn't know how to back our rig, so we had to keep going forward regardless of circumstances.

We finally arrived at the farm late in the evening, and an older couple was waiting for us. "I was worried you'd given up and gone to bed," I said.

"When you are hauling a horse, you keep going until you get there," she replied. How true! I thought of her comment all the years that I was a horse owner.

"I can't believe I left my horse with someone I don't know, and that another unheard-of person will come and drive him to Anchorage, Alaska," Bob commented as we left town early the next morning.

We arrived in Anchorage in May of 1971, and Pokey arrived at Elmendorf Air Force Base stables a week later. He survived the harsh winters in Alaska by growing more and more hair. He looked like a giant teddy bear.

Soon after school started the first year we were in Anchorage, Rob's eighth-grade science teacher wanted to have a race between his fastest student and a horse. Rob thought he must have mentioned that he owned a horse. The race was on, and the whole middle school would be let out of their classes to watch. The theory presented to the students was that a human is faster to begin with, but once the horse gets going, he can last longer and win the race.

Rob came home and laid out the plan to us, saying he would be riding Pokey in the race. He felt his reputation required him to ride since he had volunteered the horse. The stable was about three miles from Fort Richardson, where we lived and where the school was located. We didn't have a horse trailer to haul the horse to the school, so Diane volunteered to ride Pokey the three miles to the post. Before getting close enough to be seen by the students, Rob would take over. For my sanity and for safety, I walked alongside Diane and Pokey to the post.

Rob was nervously waiting at the meeting place. We could already hear noises coming from the bleachers at the school. The students were outside and waiting. Pokey could hear the noise, and he became nervous, jumping all around. We had a difficult time getting him to be still long enough to switch riders. Once Rob was in the saddle, we all started walking toward the school. Pokey whirled around and tried to run in the opposite direction as we neared the school. Suddenly the whole group of middle school kids saw us in the distance, and a loud roar went up.

This really upset Pokey, and he wasn't going any closer to the noise. In fact, he insisted on getting out of there. He'd had enough. Rob made a few passes from side to side. The noise level increased as everyone realized they could see the horse and rider. I yelled loud enough for Rob to hear me and told him it wasn't safe to continue, to turn around and head in the opposite direction from the crowd. Diane agreed to take the horse back to the stables. After we had gone out of sight from the watching students, riders were switched again. I had another three-mile hike back to the barn with Diane on a very excited Pokey. Rob went back to school to face all the comments. He wondered why he had ever mentioned having a horse. I wondered why I had ever agreed to such a scheme. It showed once again how little this mom knew about horses.

Three years later, we were faced with getting Pokey back to the lower forty-eight. We had a horse trailer and did have experience in hauling a horse. My job was to prepare the trailer to keep the dust out on the 1,400 miles of unpaved roads and to have enough supplies to get us to Northern Virginia.

It was a long, strenuous journey over the great Rocky Mountains. We went up one steep mountain, down the other side, and up again, each one getting steeper. We constantly worried if we'd be able to climb the next mountain with our heavy load. The scenery was rugged and breathtaking, but I was too worried to enjoy it. On the trip to Alaska, the Rocky Mountains hadn't been so intimidating because we didn't have the extra load of a horse, horse trailer, and all the necessities for a horse.

Fourteen days later, we arrived in Fairfax, Virginia, tired but happy to have made the trip without incident. Pokey had spent nights in Indian hunting corrals, fairgrounds, air-conditioned stables with piped-in music, campgrounds with zoo animals, and private barns. He always wanted his traveling home within his sight; otherwise, he was worried and incessantly called us.

We continued hauling Pokey from one state to the next until his last days. Bob and I acquired some experience at managing horses on the ground, but we never rode in those days. We couldn't ride and didn't have the time or money to entertain such a thought. Diane showed Pokey in horse shows, and she didn't want someone who couldn't ride to confuse her horse by trying to ride. Rob had long ago turned his half ownership in Pokey over to Diane. His interest had moved to Boy Scouts and girls.

CHAPTER 3
SEARCHING FOR MY HORSE, 1993

The search for my horse began in earnest. We started looking at ads in *The American Quarter Horse Journal* and other horse magazines and sought information from trainers and other experienced riders. It was assumed I'd get a tall, long-legged quarter horse suitable for the English riding classes. This meant riding in an English saddle like the jumpers use, instead of a western saddle with a saddle horn. I didn't think this through. In a western saddle, you can hang onto the saddle horn if you become uneasy or afraid. Getting a good English riding horse came about because Bob had Diane ride Dan in the morning English classes so he'd be warmed up for Bob's afternoon novice western pleasure class. This worked well until Bob became a better rider and needed more warm-up time on Dan. The timing seemed perfect. I couldn't ride, so Bob and Cathy thought a horse suitable for Diane to continue riding English was needed. I did want my daughter to ride and show my horse while I sat back and watched with pride. My goal was to own a good horse and learn to ride. I never thought I'd take it seriously enough to get in the show ring. The top priority for me was to have my own registered quarter horse.

An ad in a horse magazine for a quarter horse gelding that was a national winner in Hunter Under Saddle, an English riding class, caught my eye. The gelding stood sixteen and a half hands and was tall and elegant in the picture. He was dark bay in color and was stabled in Ohio. Everything in the article sounded great. A hand is four inches, and dark bay is almost black. Cathy was from Ohio. She contacted her family veterinarian to ask him if he knew a vet near where the horse was located. We would have preferred for Cathy to go check out this horse, but she was eight months pregnant and couldn't fly. The recommended vet didn't do business at this barn. Wanting an impartial evaluation, I arranged for him to go examine the horse.

"My, he is a handsome horse!" were his first words when he called after seeing the horse.

My heart leaped for joy when I heard this. My desire to have a horse seemed to be getting closer to becoming a reality.

"I have thoroughly examined the horse and drew blood to make sure he didn't have drugs in his system," the vet said. "He's been worked on a lunge line to test for free and unhindered movement by asking him to walk, trot, and canter in both directions. Do you want any further testing?"

I agreed the most common problem areas had been covered. Cathy, Bob, and I had watched a video of this horse working both English and western styles of riding, and we had been impressed. To someone wanting a horse, every detail sounded wonderful. I was thrilled and excited yet worried about taking such a big step. Decision time was near. Thinking of the price of this show horse, reality would hit, and those austere growing-up years on the farm would flash through my mind. How could I justify spending so much money on something I didn't have to have?

After much fretting and more worrying, a lesson my dad had taught me many years ago came to mind. He'd called one day when phone calls were rare and I was living halfway across the country. Rob was a baby. I'd quit work to stay home with him, and money was tight. That morning while I was pressing Bob's military shirts, my iron had stopped working. Having this on my mind, I kept bitching to my dad about the iron.

"Nell, do you have the money to buy another iron?" he finally said.

"Well, I guess so," I replied, stumbling for words.

"Then you don't have a problem, do you?"

Remembering this allowed me to temporarily stop questioning my decision and to begin moving forward.

The horse's registered name was Sonny Corleone, and his barn name was Marlon. At first, I thought he shouldn't be called Marlon. It didn't seem a suitable name for such a handsome horse. However, I didn't like Sonny because I knew two old plug horses with that name. Marlon it stayed. His father's name was James Caan, further carrying out the *Godfather* theme.

The trainer at Marlon's barn was bringing a load of horses to Lexington, Kentucky, the first weekend in June. She had room for Marlon. We made arrangements for her to include him.

On a sunny summer day in June, Bob and I left Huntsville, Alabama, and made the five-hour drive to Lexington. We marveled at the beautiful Kentucky horse country with well-kept grassy pastures that went on and on as far as the eye could see. They were filled with horses—mares with babies, yearlings, and adult horses. From a distance, they all looked to be thoroughbreds. This was Kentucky Derby territory, where these horses were raced. What an idyllic scene for a horse lover. It couldn't have been more perfect.

The tension was building as we neared Lexington. As a rule, I don't fear the unknown. This was different. My emotions ran wild. I was happy yet worried. Would this be the horse? Thankfully, it was a short night. We got in late, and by eight the next morning, we were on the showgrounds.

Horse shows start early in the morning and continue until the list of classes has been completed. How long it takes is determined by how many people have entered that day's classes. As per instructions, we found the riding arena and waited for the trainer and Marlon to arrive. Soon we spotted a lady coming toward us, leading a horse like the one we had seen in the video. Seeing the tall, dark, thoroughbred-looking quarter horse, I thought it almost unbelievable that he could become mine. Being older and not able to ride, I dreamed of owning a horse, but never such a beautiful show horse. When they stopped in front of me, I was speechless. He was just as striking and handsome as the vet had said. He was s-o-o-o tall, and his large, liquid brown eyes appeared bottomless. I just stared at him and tried to take him all in. He stood quietly and looked at me. At that moment, I became hooked on him for life. His demeanor told me he needed me like I needed him.

So much had already been invested in this horse with the pictures, the video, and the vet check that he was probably going home with me regardless. The trainer rode him around for a few minutes and then asked if we wanted to ride. I couldn't ride at all, so it was up to my brave, new-to-riding husband to try out the horse for me. This was a joke because he didn't know what to look for. The ride was considered successful if he didn't fall off. After walking around the arena for a few minutes, Bob got off the horse, looked at me, and indicated that he didn't know any more than he had before his brief ride.

"What do you want to do?" Bob asked.

"I want him!" I immediately replied.

A woman from Marlon's barn was watching ringside. She tried to warn me that Marlon had severe problems and was angry at the world. "I never knew he had ears; he kept them flat back all the time," she shared.

Looking back, I see this wasn't a good sign, but I was so thrilled to be taking this gorgeous horse home with me, I didn't care. Later, her comment would lodge in my mind, never to leave. Marlon was loaded into our horse trailer, and my new journey began. We tried giving him water en route; he just looked at us, kept very still, and was so quiet. We couldn't help worrying about him. He didn't seem to mind traveling alone or to wonder about where he was going.

Diane and her girls, Jessica and Rebecca, were at our barn when we arrived. There was much excitement to see Mama's new horse. They gathered around as we unloaded Marlon. He backed out of the horse trailer, stood looking around a few minutes, and then started grazing the green grass. We offered him water; he drank it and appeared OK as far as we could tell.

Dan was in our barn, and he was very interested in the new horse. He wanted company. The main barn had two big stalls, a tack area, a place for hay, and a hallway for grooming horses. In addition, we had added a large stall to one side of the barn. Dan was in one of the stalls in the main barn. We put Marlon into the other one. Later that day, we heard loud noises coming from the barn. We rushed down to see what was happening. The horses were just standing there looking innocent. We couldn't tell if there was a problem.

Marlon had been with us for only a couple weeks when Bob was scheduled to attend a company working weekend in Maryland, and the wives were invited. Diane and her husband, Mike, agreed to look after the horses while we were gone. We flew out on a Friday and were due back Sunday evening. On Saturday, I received a call from Diane saying there was a big hole in the wall in Marlon's stall. They couldn't figure out what had happened. The only thing Diane could come up with was that Marlon had tried to roll and kicked a hole in the stall.

"He must have lain down to roll and in the process got too close to the wall to get his legs under him to get up. While thrashing around, he put a foot through the wall," Diane told me.

She felt Marlon should be moved to the outside stall. His current stall had two twelve-foot-by-six-foot heavy rubber mats with lots of sawdust on top. If he changed stalls, the mats would have to be moved. Diane and Mike struggled most of Saturday to switch stalls for Marlon.

When we returned home, we continued to hear the loud bangs from the barn, almost like a gun going off. We didn't know why. Rushing to the barn, we still couldn't see any signs of trouble. More holes appeared in the walls, though not as huge as the first one. Later, Bob was in the barn when Marlon whammed the wall as hard as he could.

"It startled me, and I yelled so loud it frightened the horses," Bob said. "I couldn't see any reason for him to kick the wall."

We were worried Marlon would become lame in his right hind leg, his primary kicking leg. Bob scouted our area and found a place that sold half-inch- and one-inch-thick sheets of rubber. They'd cut it to any size you needed. Bob brought home a huge sheet in the pickup truck. We put it on the wall behind where Marlon usually stood. He liked to hang his head through the window-size cutout into the main barn. We couldn't keep him from kicking, but at least this would soften the blow to his leg and ankle. The kicking went on and on. We later caught him kicking other areas of the stall. Hoofprints appeared in the oddest spots. We kept putting sheets of rubber where we saw hoofprints. We ended up covering the front of the stall too, because sometimes he'd whirl around and kick that wall. His stall was now totally rubber lined in a most amusing pattern.

I finally solved the mystery of why Marlon was doing most of his kicking. I kept looking and finally found a tiny opening between boards where Dan and Marlon could barely see each other. It was just big enough for Dan to put one eye up to it and send Marlon a message that he wanted to be boss. But Marlon was number one in the pecking order, and he intended to keep his position. When the horses were out grazing, there had to be a pasture between them; they couldn't share a fence line. Biting and kicking each other, fighting for dominance, they'd knock down the fence. Ever so slowly Dan accepted that for now, Marlon was boss. He continued to test Marlon to see if he could become number one. They became buddies, but this had nothing to do with their desire to be at the top of the pecking order. Marlon never missed a chance to let Dan know he was in charge. He had to be first at everything. First to be fed, first to leave his stall—on and on it went. That never changed. In the early days, if I was in his stall, he would run into me if he suspected Bob was taking Dan out of his stall.

Sometimes Bob got tired of waiting. He'd get Dan ready to go out before I was ready, and Marlon could tell. Then I'd have chaos to deal with. Dan was easy for Bob to handle; he'd lower his head and help Bob put his halter on. Catching Marlon was the opposite, and every day differed in how long it took. If Marlon heard them getting ready, he'd become very agitated, and it would take me longer to put his halter on. He'd throw his head up in the air and start running in his stall without any place to go. Staying alert and ready to sprint out of his way was uppermost on my mind. After being knocked up against the wall a few times, I was fully aware of what could happen.

"Bob, you and Dan have to be patient and wait until Marlon and I are ready," I told Bob. It was difficult for them to wait for long periods of time.

This wasn't Marlon's only problem. Not long after we brought him home, he became colicky, and the veterinarian was called. Colic consists of severe abdominal pains, which can be critical, as horses cannot vomit. Twisting an intestine can be fatal if the horse lies down and rolls. We walked Marlon all night. The vet made three emergency trips to our barn that night to treat him. After such a severe attack, it takes time for a horse to get his strength back.

A short time later, he had the beginnings of another colic episode, and we called the vet again. Marlon was so unhappy and unruly when the vet started to examine him that the vet felt it was necessary to put on a hard hat.

"Oh no, this isn't good," I said. "I have never seen a vet put on a helmet."

It turned out to be a splendid idea, as my dear horse became angry when the vet stuck the thermometer in his rectum to take his temperature. The vet suggested we do blood work and see what the underlying problem was. The test indicated Marlon was dangerously anemic. The vet instructed us to get a gallon jug of liquid iron and to give him a dose daily with his regular food.

Marlon had a type A personality, was very hyper, and didn't trust anyone. Anytime I went into his stall or near him, he would look me up and down to determine why I was approaching him. He checked to see if I had anything in my hands, and if so, he wanted to see and smell it to determine what it was. If it wasn't to his liking, or if he didn't have a clue as to what I wanted, he'd bitterly complain. He'd watch me for a few minutes, and then the transformation from questioning to anger would start with a wrinkle just above his lips. The wrinkles moved up his nose and face in slow motion until they reached his ears, and then the ears went flat back. This process was beyond belief. When those ears went back, I made sure I was out of his way. Anything could happen then. If this didn't scare me away, he'd try other maneuvers like racing around and around in his stall, slinging his body and head, and leaping into the air.

I'd move outside the closed stall door but wouldn't leave. He could see me standing quietly, acting as if nothing was happening. When he realized I wasn't leaving, he'd calm down a little. Then I'd open the door, step inside, and walk toward him, talking in a soft, soothing voice. Instinct told me he wanted to trust me. We heard rumors that Marlon had been mistreated by some of his past trainers, and it was evident that he was mad at the world. He was a smart horse, and he wanted to understand what was being asked of him. When he was mistreated, he fought back. So many horses are too scared to fight back. Marlon was the opposite, and he never gave up as long as he had the strength. At first, he was so quiet, stood so still, and just stared at me. At the time, I didn't know why. With good food and the iron supplement, he began to feel better, but we were in for another unbelievable surprise. All hell broke loose, and he became meaner by the day.

"As each drop of iron enters his bloodstream, he takes out his frustrations and anger on us," I said.

Before Marlon, Diane's husband, Mike, had hung a rope from the ceiling in the barn hallway for Rebecca to swing on. Marlon considered it a personal affront for her to swing on the rope. He didn't like kids, period. They were always moving around, and he couldn't predict what they would do. He especially didn't like her swinging on that rope; it was near the half window where he hung his head over into the main barn. If he saw Rebecca, he showed her his threatening look. Quick as a flash, the ears would go back, and all the others tactics he employed would show up. If this didn't scare her away, he would shake his head from side to side and add kicks to the wall with all his strength. He terrified Rebecca. It was enough to put fear into any mortal. If you were nearby when he was out of control, he wasn't above biting you. This was my new horse, the one I had painstakingly chosen. Marlon was supposed to be an older, calmer horse with more training under his belt, and a winner in the show ring.

Our daughter was an excellent rider. We had hauled her horse hundreds of miles so she could ride and show. We helped her with her horses and learned many tricks horses love to pull on people who know absolutely nothing about them. If I hadn't been exposed to lots of safety rules for handling a horse on the ground, Marlon would have severely hurt me.

"My number-one rule is to get out of his way when he is upset and angry," I'd laugh and say when questioned about my safety.

However flippant my reply, I took the rules I'd learned from Diane's horses very seriously. They were uppermost in my mind as I dealt with Marlon. After suffering a sore hand, I knew never to wrap a lead rope around my hand when holding a horse. Always walking by the head when leading a horse is an important rule to remember to prevent him from stepping on your toes. Wear hard-toe boots, just in case. Finish what you start. Don't hold your daughter's horse by sitting on it, even if she orders you to. When I did this once, the horse was facing an open stall door, and he started walking through it. As my face was meeting the top of the door frame, I bent backwards and pulled a muscle. It was painful but better than a full-face hit. Don't startle a horse; speak in a calm voice when you are approaching it, or you might get run over. It is always good to hold out your hands and let a strange horse smell you and get acquainted. These are good practices, and I tried to remember them all.

In all those years of hanging around barns where we boarded Diane's horses, I had never seen a horse as out of control as Marlon. I thought Diane would know how to handle him. She didn't mind riding him, but she didn't want any part of him on the ground. He threatened her continually and was so intimidating that she didn't want to deal with him. The message was clear: I was the owner, and the problems were mine. He terrified me too. He learned that I hung in there until we had a resolution of some kind, even if it wasn't perfect. I'd open his stall door with his halter and lead rope in my hand. He'd see me and immediately warn me against coming into his space. If this didn't work, he'd try all the threatening moves in his repertoire. The scariest one for me was when he whirled around and kicked in my direction. He kicked whether he had a reason to or not, and my presence in his stall was reason enough. Frustration and fright were my normal feelings in dealing with Marlon. But quitting wasn't an option for me.

Besides worrying about being kicked, I had to worry about being bitten. When I walked around to his head, he'd turn faster than I could walk. This would go on and on. I'd scold him now and then. Sometimes he'd throw in a good loud kick to the wall for extra emphasis. Catching him took a long time, and when he was at the top of his game, it took even longer. My training in handling an angry, out-of-control horse was limited, but I stayed until he was caught. When I finally snagged him in a slower bypass or got close to his head, I'd talk soothingly and caress him while carefully watching to make sure he didn't bite me. Early on, I discovered that if I got close enough to quickly put the lead rope around his neck, this would slow him down enough to let me put his halter on. Later, I tried stepping just inside his stall door, holding out the halter for him to see. Then I waited. I'd been told that eye contact conveyed a challenge. In case this was true, I dared not look him in the eye. The wait was long. It seemed I stood there forever, staring at the sawdust and watching him out of the corner of my eye.

"The time you're waiting doesn't seem to be getting any shorter," Bob often said.

Later, it wouldn't take as long for Marlon to calm down and let me approach him. At those times, I felt lucky. If it was time to go out to graze, he'd be more cooperative. When I needed to put a blanket on him, which he bitterly opposed, or was catching him for some other reason, it was an entirely different story. To say he wasn't fond of blankets and hoods was an understatement. Show horses wear blankets and hoods in cold weather to keep them warm and prevent them from growing winter hair. They need their beautiful summer coats of hair for the show ring. Also, less hair makes riding easier in the winter since the cooling-down period is shorter after a workout.

CHAPTER 4
ADJUSTING TO LIFE-ALTERING EVENTS

I delayed learning to ride Marlon, giving him time to calm down and adjust to a new owner. After he had been with us for over a year I felt it was time for me to get started. I don't think about what is going to happen in the future until it is written in stone. When life-altering events do happen, I can be overcome with emotions. Facing these events can take days to adjust to, regardless of whether the event is large, small, truly wonderful, or terrifying. It takes over my mind, fills my heart to overflowing, and just envelops me. I am so excited or devastated that physical action has to be taken, like going for a jog after hearing the news that a grandchild has arrived. I often did slow walks around my pastures (with lots of tears) upon hearing sad, heartbreaking news such as the death of a friend. Later I wondered why I didn't let myself read the obvious signs and prepare emotionally for what was coming. But I gave up on trying to change such a deep-rooted trait of mine. I have always believed you shouldn't stress yourself until necessary. If you spend time, energy, and emotions on something that might never happen, then it is for naught. However, when things do happen, you are unprepared, and you ask yourself over and over, "Is this for real?"

My first major life-altering event was when I went away to college in the fall of 1951. It was the first time I had left my small, secure area and entered the big world, and I was alone. When I say alone, I mean *totally alone*. I had never seen a college campus, didn't know anyone, and couldn't communicate with my family or friends except by mail. That didn't help when a crisis occurred and a kind word was needed. I can still feel the loneliness, homesickness, and doubt about whether I could survive it almost as if it is occurring again. As my painful emotional state abated, I found a way to adjust. It was one of the hardest things I ever had to do.

Daddy always encouraged us to make good grades in preparation for attending college. He helped us with homework in math, the science courses, and history, but he had nothing to do with English. He dismissed it and said it was unimportant. Daddy kept me ahead of the class in math, and my classmates thought I was smart.

Finishing junior high and starting at a rural county high school, I discovered that algebra was the only math course taught other than business math, if that counted. In my dad's opinion, it didn't. I took algebra, and we only covered a few chapters all year. The course did help me come to terms with using letters in math, as this was such a foreign concept.

"You can't add a's and b's," my classmates and I would say to each other. It took a long time for this to become normal to us.

The next year I took chemistry, and I excelled in this class. Being an up-and-coming high school senior, I was concerned that the courses offered at my high school wouldn't prepare me for college.

"Why don't I go on to college and take the noncredit courses that I will need?" I asked. I was astounded when my dad, and even my chemistry teacher, began to take me seriously. Why, I'll never know. I didn't have a clue about what I was proposing to do.

We lived out in the country in a very isolated area, and we didn't have television. Word that there was such an invention had just trickled down. It meant nothing to us at the time. Some of the local people didn't believe such an invention was possible. They kept saying, "How can you send pictures over electrical lines or through the air?" It was highly debated.

In my small world, there weren't any foreigners, Jewish people, black people, or Yankees. I had seen a few black people in town, but I had never spoken to one. The Catholic faith was mentioned occasionally. I did not know anyone who was Catholic. On the surface, all the people I knew were the same as me.

Wanting to chart my own course, I dreamed about a career in chemistry. Dreaming was free, so I indulged in it and explored all possibilities I'd ever heard of. I consulted with my chemistry teacher, and she gave me information and addresses to send for catalogs from four colleges in Alabama.

Daddy and I pored over the catalogs. I couldn't imagine what a college campus looked like or what you did once you got there. I just knew I had to get there and see for myself!

Applications were sent to all four colleges, and information about jobs and scholarships was gathered. Work scholarships were offered to me at all the schools. Later, I decided that a work scholarship meant the school officials thought you could work thirty hours a week and make passing grades.

Mother kept insisting the girls' college was the one for me. I was young and very unworldly. At the girls' school, they dressed for dinner and had one formal evening each week. In my heart, I knew this wasn't a school for a country girl like me, with barely enough clothes to get by.

Finally, my dad made one of his bold statements, saying, "I think sending a girl to a girls' school makes her boy crazy."

I hadn't thought of that, but it sounded good to me. It was settled that I wouldn't be going to the girls' school.

In late spring, I officially applied to the University of Alabama. The university accepted me with the caveat that I needed to take two noncredit math courses. My scholarship job, as they called it, was to work thirty hours a week in the girls' dining hall. It served three meals a day, seven days a week, holidays included, and I'd work every meal except for one weekend meal a month. Uniforms were required for work, which meant changing clothes before and after each meal, three times a day. Later, when I learned how tough my schedule was, I was so thankful I hadn't known beforehand.

Looking forward to all the possibilities that my new adventure entailed, I was excited. The panic and self-doubt would come later.

The big day arrived for me to leave home. My dad worked at a construction job in Chattanooga, Tennessee, approximately fifty miles away. He couldn't miss work, so he had hired Willie Bee, a friend, to take Mother and me to the university. It was a four-to-five-hour drive from my house, so Willie Bee arrived before Daddy left for work between four and four thirty in the morning.

Standing in our living room at four in the morning and saying goodbye to my dad, brother, baby sister, and everything familiar to me was overwhelming. As I faced leaving home and heading into the unknown world, my bravado faded, and emotions took over. We hugged, cried, and then we did it over and over. Even Willie Bee shed a few tears. I imagined this was what it felt like to go off to war!

"If you aren't finished studying by midnight, go to bed anyway, because if you aren't rested and fresh the next day, you haven't gained anything. You are barely seventeen years old, so leave the boys alone this first year and get yourself established." That was Daddy's advice that morning.

The funny thing is that I kept thinking of this advice all through my college years. I always gave myself permission to go to bed by midnight. In the beginning, there wasn't time for boys, so they weren't on my list of things to do or not do.

When we arrived on campus around noon, the first thing we saw was a large crowd of dressed-up girls squealing, laughing, and having a great time. I could not imagine what was happening. The catalog hadn't mentioned this, and I had read every line. We plodded on to find my dorm. We were welcomed by the vivacious older house mother, who guided us to my room. Mother and I took my suitcase and my box of linens up to my room, unpacked my few things, and made my bed.

It wasn't long before Mother and Willie Bee began to worry about the long drive home. The time had come, and the pain of this moment, which had been buried deep inside me for so long, surfaced and threatened to overwhelm me. I could not stand the thought of them leaving, yet I knew they had to. After the tearful goodbye, I was in shock and just felt numb. As they drove off, waving goodbye, I realized how alone I was even among a crowd of people. Before dwelling too much on my misery, I had to find my place of work and report in. I checked my campus map to find my way to the dining hall. Walking the shaded sidewalks to get there, I had to pass the football players' dorms and workout buildings. Benches were located under large shade trees near the sidewalk, and there sat several big burly guys watching as I approached. They started making comments in a non-Southern accent before I got there, as I was passing by, and as I continued onward. That day, I was in my own lonely world, and it hardly fazed me. This continued for my entire freshman year, but I adjusted to it as time passed. I came to realize that they couldn't resist teasing this freshman girl going by in her pea-green uniform several times a day, every day.

A few days after my arrival on campus, registration for classes began. Classes were offered on Monday, Wednesday, and Friday, or Tuesday, Thursday, and Saturday. By the time I got off work and arrived at the auditorium, most preferred class times were full. I ended up having four Tuesday, Thursday, and Saturday morning classes starting at eight o'clock. My advisor didn't do me any favors; he advised me to take a full load of classes with two three-hour labs, even though I worked thirty hours a week.

The only communication with my family was by mail. After receiving homesick, sad, and pitiful letters for a couple of weeks, my parents drove twenty miles to town after working all day and called the hallway phone in my dorm.

Someone yelled my name, and I was surprised to have a phone call. Hearing Daddy's voice, I started crying.

"Nell, you don't have to stay. I will come and get you this weekend. You cannot make yourself sick. Nothing in life is worth that," he said.

"No, I don't want to come home. I just want to feel better."

They managed to have enough change for my mother to say a few words, and then they were gone. Standing there, it felt like a dream had occurred. In my misery of feeling so alone, I now knew my parents had been right there with me. I realized how much they cared about me and my situation.

Because of time and money, I wasn't planning on going home until Thanksgiving. Four weeks after the phone call, my parents insisted I come home for a short weekend visit. This started me on the road to recovery. I began to feel better day by day, but I continued to struggle as I adjusted to my hectic schedule. There was never enough time for studying. I was always having to decide which course was most important to tackle. After struggling along for three semesters, I had to add a class in scientific German. The objective was to translate the German science manuals into English. This was the most difficult course I ever had to take. It proved to be too much to handle with my already loaded schedule. The next semester, I found a job as a quantitative analysis lab instructor, working two three-hour labs a week. This lowered my stress level, as it lightened my workload, and I had more time to study.

Chemists were in demand, and companies came to the college each fall and offered career jobs to the seniors. To interest us in their companies, they offered summer jobs throughout our college years.

School began the day after Labor Day at the University of Alabama in 1953. I had just returned from my first summer job with Dupont in Chattanooga, Tennessee. I was starting the second half of my junior year and the second semester in my new job as a quantitative analysis lab instructor. My lab was made up of twenty-five mostly male students. All were older than I was, but I never gave it a thought until I met Bob.

A second quantitative analysis lab with a different lab instructor was directly behind my lab. Those students had to walk through my lab to get to theirs. A tall, handsome guy with curly black hair and dark brown eyes was always around being charming and cool anytime I had a spare minute. He was in the back lab, and this gave him extra opportunities to be noticed as he strolled through my lab. As the busy days flew by, I did take time to notice he had a quick wit and a wonderful sense of humor. His name was Bob, and soon he asked me for a get-acquainted date at a popular hangout on campus. He didn't have a car; very few students did. In talking that evening, he told me he was a senior and twenty-two years old. Because I was the lab instructor, he assumed I was older, and he had added a year to his age. When he found out I was only nineteen, it was too soon to admit he had told me an untruth.

Bob appeared before my classes, after a class or lab, and even in the library to help me study. We spent evenings together getting acquainted and discussing our dreams and hopes for the future. I should have been studying, with all my afternoons filled with labs and lots of early-morning classes. Occasionally, when I was crossing the quadrangle early in the morning, Bob would appear, fall in step with me, and walk me to class. It was unusual for him to be up this early, as he never scheduled an early-morning class if he had a choice.

When we were ready to leave school for the Thanksgiving holiday, Bob asked me, "Will you stop by my house to meet my mother?"

Bob lived in Bessemer, Alabama, and the Trailways bus that passed in front of my home in Henagar, Alabama, had a long layover in Bessemer. A girlfriend of mine, also from Henagar, was with me. Bob met us, and we rode the city bus to his house.

"You should choose the other girl because she is taller," Bob's mother told him after meeting us. That was her only comment.

Back at the bus station, Bob said, "How about I ride the train up Sunday morning, spend the day, and meet your family? Late afternoon, we can catch the train in Fort Payne and ride back to Tuscaloosa together."

I was concerned about his visit; this was the guy my dad had told me to forget a few weeks before. The visit was tense for all of us, but my dad did behave himself. On the train that evening as we were talking, we felt the visit had gone as well as it could under the circumstances.

A week later, I was again crossing the quadrangle for an early-morning class when Bob stepped out from the shadow of a building. "Why are you up so early on a Monday morning?" I asked him. He seemed nervous.

He walked closer, put his arms around me, and said, "Will you marry me?"

Not being a person to anticipate what might happen until it does, I was stunned! I was happy, excited, and loved every minute I spent with Bob, but I was constantly stressed because of my heavy class load and lack of time. However, it took only a moment for me to recover.

"Yes! When?"

"This weekend, next weekend, or the week after."

"OK, this weekend."

We started making plans for a license and then a trip across the state line to Columbus, Mississippi, on Saturday. Mississippi didn't have a waiting period. Midweek Bob appeared again early one morning, needing to talk to me.

"What are your thoughts about waiting another week or so to get married?" he asked.

"It is now or never," I quickly replied, without giving it a thought.

In the world I came from, your word was everything. Feeling he had made a commitment, I was determined that he would keep it, or I wouldn't marry him. Also, I knew that our circumstances wouldn't change anytime soon. He immediately reassured me the wedding would go on as planned. Bob arranged for his best friend to take us, along with my best friend, to Columbus the following Saturday, December 5, 1953.

Just before we left, Bob said, "I want to set the record straight on how old I am and explain why I added a year to my age. You were the instructor, and I thought you must be older."

We traveled to Columbus, found a large Baptist church, located the pastor, who was in his office on this Saturday morning, and asked him to marry us. After talking with us for a long time and asking many questions, he married us in the beautiful church parlor. We went to lunch and celebrated our wedding before heading back across the state line. Bob's friend drove us to a motel near campus. It had individual cabins, restaurants nearby, and was on the bus line, so we could get to our classes on Monday morning. We were in love, we were married, and we could temporarily forget the family problems. We did accept that we might be totally on our own, and we were ready if necessary.

Earlier in the fall, I had written several letters to my family telling them about Bob. After the third one, I received a stern letter from my dad.

"You're in college to get an education and not to worry about boys," he wrote.

That was on my mind as I wrote and mailed a letter to my parents on Saturday morning, telling them we were getting married that day.

The letter was delivered at noon on Monday. My five-year-old sister and an aunt were present when Mother opened the letter. She started reading, immediately started crying, and told the others, "Nell got married!"

Bobby, my brother, came in, and saw them holding my letter and crying. He thought something terrible had happened to me. When Mother told him the news, he was ostracized when he exclaimed, "Isn't that good news?"

When my dad came in for lunch, he read the letter, and they all cried together. They had liked Bob well enough, but they weren't ready for their oldest child to get married before graduating from college. The Depression had made it impossible for my dad to go to college, but he helped make the dream come true for me, even though we faced many hardships in doing so.

After our wedding, Bob's friend was going home to Bessemer, and Bob asked him to go by and inform his mother that her only son was married. The report back on Sunday evening wasn't good.

The following weekend Bob went home without me to face the music. It was much worse at his house. His mother threw what a southerner would call a "hissy fit." She ended up telling him, "According to the Bible, the only son can't get married. He has to look after his mother." She hollered, screamed, and threw things around, and they battled all weekend.

Bob and I hoped our parents would continue to help us financially through the current school year. We had jobs, which paid most of the expenses, but they didn't cover everything. We knew our chances would be better if we told them after we were married and let the chips fall where they may. It would be too late to make threats that they felt they had to keep.

My dad missed his opportunity for compromising when he answered my letters, telling me to forget Bob. Later, when he asked why I didn't come home to get married, he knew the answer. Discussing marriage before finishing school wasn't an option. I'd tried introducing Bob as my first serious boyfriend, and it had been a disaster.

Two weeks after getting married, we went home for the Christmas holidays. We spent the first week with Bob's parents, giving me an opportunity to meet his dad and three married sisters. Also, Bob wanted to work as many days as possible at his brother-in-law's electrical repair shop. He'd been working there on Saturdays and as needed since high school. Getting ready to go to work, he said he needed a couple of peanut butter and jelly sandwiches for lunch. Mrs. Parsons set the needed supplies on the countertop, preparing to make the sandwiches. I quickly volunteered. I put the peanut butter on a slice of bread, then tried to spread the jelly on top. I struggled with it, not making any progress and tearing up the bread. Mrs. Parsons was watching me.

"Why don't you put the peanut butter on one slice of bread and the jelly on the other slice, then join them together?" she finally asked.

I was so embarrassed. Thankfully, she didn't make a big deal about it. I had never made nor eaten a peanut butter and jelly sandwich.

One morning as Bob was leaving for work, he said, "Come uptown over my lunch hour. I have a surprise for you."

I rode the city bus and found my way to the repair shop. He took me next door to a pawn shop, which was owned by a family member of his brother-in-law. Bob had negotiated for a wedding ring for me. He had picked it out but needed to check the fit. With all the family turmoil we were going through, a ring hadn't crossed my mind. I was so proud of my new husband for thinking of a ring and wanting me to have one. It was beautiful to me, a band with teeny tiny diamonds across the top.

The second week of Christmas break, we caught the bus to my house. My family and friends had hastily arranged a beautiful wedding reception and shower for us. I got to see everyone and introduce them to my new husband. We received dishes, glasses, silverware, assorted towels, two sets of used double-size bedsheets, a used cast-iron skillet, and a few pots and pans, including a coffeepot. The gifts gave us the necessities to set up housekeeping in a furnished efficiency apartment. Reaching into their own households, the family added anything we needed that we didn't get. I felt such love for my large, extended family and shed tears of joy mixed with thankfulness and appreciation, knowing they had sacrificed to give us these extra things. We packed our gifts in boxes so we could take them with us on the bus back to school. In early December, we had checked out efficiency apartments. Bob had put a deposit on one for January when we returned.

Having only visited my home once, Bob was unprepared for the country ways on Sand Mountain. He felt he was constantly trying to solve a mystery when confronted with the local customs. He and my brother were sent to the Henagar school principal's house to borrow folding chairs for the shower and reception. Driving my dad's pickup truck, Bobby parked along the road in front of the house. He got out, walked a few steps into the yard, and hollered, "Hello!" as loud as he could. He kept hollering until someone came out onto the porch to find out what he wanted.

Returning, Bob couldn't wait to tell me what had happened. Bobby's behavior was unbelievable to him. "Why didn't he walk up and knock on the door?"

"Most country houses have dogs lurking under the front porches, and they don't take kindly to an unknown person walking up and knocking on the door. That's a good way to get dog bit," I explained.

Although my dad hadn't wanted me to get married until I graduated, he was kind and welcoming to Bob in his country way. One afternoon, they had a long conversation.

"Talked to your dad for an hour about someone being paroled before I discovered he was talking about a payroll," Bob told me afterward. I did a lot of interpreting in those early days.

Back on campus in early January, we happily moved into our first home and started our life together. The apartment was barely adequate, but we didn't care. We were more in love than ever. The apartment was older, dull in color, but appeared to be clean. Bob checked out the small gas stove, the dilapidated refrigerator, the running water to make sure we had both hot and cold, and the bathroom.

"Everything works," he proclaimed as I was putting our clothes away.

We made up the bed with the used sheets and put the quilt my grandmother had given me years ago on top as our bedspread. The pattern of the quilt was the little Dutch girl wearing different-colored costumes. It was beautiful! It truly lit up the room and reminded me of how much I was loved. My grandmother had worked long hours making this quilt for her oldest grandchild.

The mattress on the bed sank in the center, and I constantly tried to keep from rolling down the incline. Weighing more than I, Bob had a more level sleeping area. It didn't take long before I had backaches. Flipping the mattress didn't help. We stuffed every extra thing we owned under the mattress and couldn't tell any difference. Next, we got cardboard boxes from the grocery store and cut and flattened them to fit under the low places, and this greatly improved the problem. The bed doubled as a couch, a desk, and the study area. Besides the double bed, there were a couple of uncomfortable, straight-backed chairs and a faded black chest of drawers in the room. All the furniture was old, used, and very beat up on the outside, but sturdy. The apartment also included a small eat-in kitchen and a bathroom. Both were old and worn, but everything worked, as Bob had determined.

I quickly realized I had another problem: I didn't know how to cook. Going to the grocery store to shop was a nightmare. Walking from one end of the store to the other, I couldn't see one thing I could use. At the time, I was too shaken up to notice the assorted canned goods. Growing up on a farm, we raised and preserved all our food, buying flour and sugar but not much else. My new husband lived in a city and had shopped in grocery stores, so he went with me. After getting the groceries home, I wasn't sure what to do with them. Bob knew a little bit about cooking, and we survived. We had bacon and eggs morning, noon, or evening. Once we tried cooking calf's liver. We were desperate for something different to cook, and Bob remembered liking it when his mom fixed it. The smell about did us in before it was fully cooked. We never, ever cooked liver again. By accident, I discovered chicken pot pies as I prowled the grocery store. I just popped them in the oven, and dinner was ready. This was a dinner I could handle. We had them so often, Bob finally said, "If I smell chicken pot pie again, I'll get sick!" It burst my bubble, and I was back to square one. They sure didn't affect me that way. I could have eaten them forever.

We'd barely figured out the grocery shopping and cooking chores when it was time to start studying for finals. Neither of us was ready, but we had to cram and try to salvage as many courses as possible.

As I took each final, I told the professors about my name change. My math professor had been my teacher for an earlier class. He was an older gentleman from Quebec, with an adorable accent, a great sense of humor, and a friendly demeanor. After class, he had often walked in my direction and wanted to speak to me. I resisted talking to him, as I was barely getting my assignments in, and sometimes they were late, my test scores were not up to par, and I no longer said anything in class.

"That explains everything, Miss Rains," he said after hearing I was married. "I have been worried about you this semester. This wasn't like you." He congratulated me and wished me a long, happy marriage.

CHAPTER 5
HOW MARLON CHANGED MY LIFE

Becoming a horse owner changed my life in every way. I am still surprised that this was true. After being a parent of young horse riders, I thought I was prepared to have my own horse, but I found it to be a different world. Having chosen an older, more experienced horse, I expected smooth sailing, but this was not to be. Trying to understand my new horse and care for him was so time consuming that most other activities were dropped. Because of Marlon's colic and anger problems, much of my time was spent nursing him or helping to repair the damage he did to our barn and fences.

Planting a vegetable garden after discovering there wasn't time to gather and cook the vegetables was a mistake. The beautiful, colorful flowers I put out across the front of my house each spring disappeared, and shrubs that could take care of themselves were planted.

Bob was taking Dan to horse shows by the time Marlon came into my life, and he was so unstable if Dan left home that Marlon had to go too. This took lots of time to prepare for, even though Diane seldom entered him in a class.

When our good friends Konny and John or other friends called and invited us to an event or an evening out, I'd often have to say, "Sorry, we are going to a horse show this weekend."

"I have never been replaced by a horse before!" Konny would comment.

Trying to understand Marlon's behavior kept me mentally weary. Taking Rob and Diane to the stables where their horses were boarded gave me lots of time to observe all kinds of horses. I thought I had seen it all. But the combination of abuse with Marlon's high-strung personality made him behave in a way that I'd never witnessed. I watched him and tried to figure out what he was thinking. I had patience with him, but some behaviors I couldn't tolerate for my safety, his own good, or our future relationship. I walked a fine line in dealing with him.

Knowing what made Marlon happy, I gave him plenty of good grain, hay, and grass to graze. Some days, I could even see small changes in his demeanor. It was scary to think about riding him, yet I knew if I wasn't older, there wouldn't be time or money for a horse. This was a trade-off that I was willing to make. When he was restless, I walked him around and tried to calm him. It was time consuming but worth it. I got insight into his real personality. He was a tall, strong, above-thousand-pound horse, and yet he let this slender older lady lead him around in open areas. He could have run away at any moment, but if it crossed his mind, he didn't act on it. I was amazed and hopeful.

I'd seen the barn cat leap up on the divider next to Marlon's face. If he was waiting to be fed, he'd frown, shake his head, and then roughly push the cat off. If nothing was going on, he'd let the cat rub back and forth along his jawline and smile while they did it. Sometimes I walked into the barn and found the orange tabby cat lying on his back while Marlon walked around the stall. I laughed and celebrated. He did have a kind heart toward critters!

He was playful with the pony at Cathy's. On his day to join the pony, he'd walk into the corral with a slight frown on his face and lower his head while gently moving it from side to side. Marlon was letting the pony know he was ready to play, but he'd better play nice, or there would be consequences. The pony was rambunctious when he wanted to scuffle. I admired Marlon for setting the rules at the beginning. After playing and chasing each other around, Marlon and the pony would stand with some part of their bodies touching.

On a warm summer day, I went to the barn to let Marlon out to graze. Walking up to his door with the halter in my hand, I noticed the half door in back of his stall had fallen down. The window on the top half of the door had been removed for better air circulation. I just stood and stared, not believing he hadn't walked out to graze on his own. He lived to go out to pasture, and he demanded to be let out on time every day. Standing there, I realized that he was in control in this situation, and he wouldn't do things he knew could get him into trouble. Any horse I've ever known would have celebrated such an opportunity and suffered the consequences later. He wanted to obey the rules when he was clear about what they were. I was astonished and happy.

As time went on, I was laughing more. He exhibited funny, ever-changing expressions when he was trying to guess what I wanted. We could relax occasionally as we got to know each other. I was grateful. Decision time had arrived.

I didn't want to wonder anymore what it would be like to ride him. I knew it would be difficult and that I would be scared, but so would Marlon. We were on this long road together.

"I have procrastinated long enough; today is the day to climb on my beautiful horse," I'd say. But things would happen, and the inevitable would be delayed again. This wasn't like me. I had faced heartbreaking, life-changing events head on, time after time. I couldn't help thinking about past difficult situations in my life. It became clear to me that I wasn't given a choice whether to face most events in my life. The only choice was to deal with them in the best way I knew how. This time there was a choice: to go forward or to abandon my dream. The latter wasn't an option I could live with!

CHAPTER 6
BECOMING PARENTS, 1957

A huge life-altering event occurred in our fourth year of marriage when I became pregnant. We hadn't discussed having children. Our marriage had included so many big adjustments that becoming a mother hadn't crossed my mind. Dealing with our families, especially Bob's mother after our elopement while we were still in college, and finding better jobs to support ourselves, kept us occupied. Another change was moving out of state one year after we married, when Bob entered the army. Being a number-one worrier, Bob had thought about children, but he didn't say anything.

"I never knew until now whether you wanted children," Bob said after he saw how thrilled I was.

The pregnancy went well, and I continued to work until the final four weeks. Also, I attended prenatal classes at the military hospital where I would have my baby. The hospital had stringent rules, and they were spelled out to us. The first rule they lectured us on was not to come to the hospital until labor pains were five minutes apart. If you did, you would be sent home immediately. The next rule stated that you and your baby would spend five days together, and you would care for the baby. The staff was there for advice and to get anything you needed, but you would be responsible for routine care. Only dads could visit for two hours each evening. The rooms didn't have telephones; there wouldn't be any outside disturbances to bother you. Prenatal classes were taught by a stern older nurse who demonstrated with a life-size baby doll how to hold, bathe, diaper, and feed a baby. I didn't know anything about looking after a baby except what was taught in class. We spent class after class practicing her instructions. She lectured us to be tough during labor and delivery and to not be a sissy.

In college, I saw a video of a woman delivering a baby. Knowing the basics didn't make me able to apply them to myself. After becoming pregnant, my curiosity kicked in early. I wanted to know if it was a boy or a girl and how it would get out of me into this world. I had many months to ponder these two questions before they were answered.

With my labor pains increasing, the hours dragging by, and heavy rain pounding the roof on a stormy night in El Paso, Texas, Bob rushed me to the hospital. I argued with him that it wasn't time, but he had this deep-seated fear that he would have to deliver this baby, and he didn't have a clue what to do. After checking me out, the attending doctor gave Bob a military scolding.

"I should send her home, but it is such a rough night, I am going to admit her," he said.

Bob was told to go home and that he'd be notified when to return. He was a nervous wreck yet relieved that he had gotten me to the hospital in a timely manner. I was hurting and fearful, but I had experts to answer questions and help me get relief from the pain. As time went by, I was reminded of another hospital rule: minimal medications are given during labor and childbirth for the protection of the baby. I was assigned a room on the labor floor. Nurses came by sporadically to check me and to ask if anything was needed. They couldn't give me what I really needed and wanted. It turned out to be a long, lonely night.

The next morning, a woman on the ward started screaming. Footsteps rushed up and down the hallway, and many voices shouted commands. The hollering went on and on. Finally, the screaming woman and her entourage came down the hallway, past my room, and went through the big double doors into the delivery ward. This terrified me. How much worse would my pain become? I lay in my bed scared and pondering what could have caused her to holler as loud as she could. The next time the nurse came in, I posed that question to her.

"Oh, she is a big baby. She should be ashamed."

To me, this wasn't a satisfactory answer. "Is this her first baby?" I asked.

"No, it's her fifth," the nurse replied with a disgusted look.

My nervousness increased as I thought about what lay ahead for me. The day wore on, and time passed more and more slowly as the labor pains picked up in frequency and intensity. I kept wondering when the final stages of labor would come. The on-duty doctor assigned to me appeared now and then. Each time I was hopeful he would say it was time to go to the delivery room and that I could be given a saddle block. The class nurse had explained the saddle block was a spinal anesthesia to be injected when the baby started down the birth canal, to help with extra end-stage pain such as tearing and stitches. Besides easing my pain, it would assure me the end was near when it was administered.

Late in the day, I was encouraged when the doctor casually told me, "You are almost ready."

"You are fully dilated and ready to go," he told me on a later visit.

I thought they should whip me onto the gurney and rush me through those double doors into the delivery room. The doctor didn't seem in a hurry.

"Your husband has been in the waiting room since early morning, and they are keeping him informed," the nurse said when she came by to ease my frayed nerves.

In their own time, they moved me onto a stretcher, and we finally went through those double doors. By now, the pressure and pains were constant and unbearable. Still, the staff took their time getting ready for the anesthesiologist to administer the saddle block. They kept uttering reassuring words to this new mom-to-be. I just wanted them to do their job and ease my pain.

Not long after the saddle block was administered, someone said, "The baby is coming! Just a few more pushes! Here he is—and you have a baby boy!"

They kept a one-sided commentary going about how great it was to have a boy, especially the first time. There would be time for a girl. On and on they chattered. I was totally exhausted, elated, and disturbed all at the same time. I had noticed the nurse quickly move the baby out of my line of sight. I had overheard the doctor tell her to do so, in a low voice with a one-word command. In that moment, I knew something was wrong.

"I want to see him," I said, weary and anxious.

"He is all bloody and messy, and as soon as he is cleaned up, we will bring him to you."

"I want to see my baby now!"

In a vague way, they explained my baby had a slight problem. I did not listen to anything they were rambling on about. I kept demanding they bring him to me immediately. Realizing they couldn't delay letting me see him, the nurse brought him to me.

Seeing him, I was gripped by fear and panic. "Can he eat? Will he die?"

He had a cleft palate and a double cleft lip. The upper lip had a double cleft, making the nose and mouth appear out of shape. With a cleft palate, the roof of the mouth is open and does not separate the mouth from the nose. Having no prior knowledge of cleft palates, I had difficulty seeing how he could survive. They all talked at once, assuring me he could eat and that with surgery he would do well. More panic set in. How could my darling, teeny baby survive surgery? He weighed less than seven pounds. I felt such apprehension and agonizing sadness. It seemed too much to endure.

"Dear Lord, why can't I take his place?" I kept thinking. "I am an adult, and I can handle it."

I don't know how they explained the problem to Bob. He was told his wife was very upset. As they whisked me to the recovery room to get over the effects of the saddle block, we were allowed a few minutes in the hallway. With stricken looks and many tears, we hugged and knew we would do everything in our power to help our tiny newborn son. Bob was told he could see me when they transferred me to a room, if he wanted to hang around. Otherwise, he would have to wait for regular visiting hours the next evening.

In a few hours, they transported me to the new-mother ward. Bob had waited, and we had another short visit together. We were both worried sick. He tried to encourage me as we compared the information each of us had been given. We had long ago decided that if we had a son, we would name him Robert the third, after Bob and his dad. He would be called Robby or Rob. Despite our troubles, we could manage a smile talking about and calling him by his name. Too soon, a nurse came to tell Bob his visiting time was over and that the baby would be brought to me as soon as he left.

It was late at night by now. Robby had been born at five thirty in the evening. I'd spent a few hours in recovery, and Bob had visited. I needed rest and sleep, but I knew the hospital rule was that mothers and babies spend the first five days alone, with the mother caring for the baby. I wanted to see him, hold him, examine him, and make sure he could eat, even though I didn't know how to feed him. A nurse brought Robby and a bottle of formula to me, saying he was hungry. She left, and we were on our own. This tiny baby and his new mom were on their own! It became an all-consuming task. He couldn't suck, so the milk had to slowly trickle down his throat in just the right amount. The end of the nipple clogged continually, and he couldn't get any milk. I buzzed the desk, and a nurse came with sharp scissors and cut a hole in the end of the nipple, trying to solve the problem. It continued to clog and was a constant problem. When the milk was flowing, he swallowed so much air, he had to be burped and shifted around constantly. As soon as one feeding ended, it was time for the next one. Robby took short little naps when he could. I didn't get any sleep night or day. It was a frustrating and terror-filled process.

The staff didn't seem to know any more than I. By the second day without any guidance, I was so stressed from lack of sleep and worry that I almost took drastic action. It was all I could do to keep from picking up my baby, walking out the hospital door, and going home. At least there would be emotional support and someone to keep me company. Fear overrode my leaving because I knew he'd need surgery, and I was afraid the army might refuse to help us if I left. Back then, we thought the military had complete control.

In the evenings, when Bob arrived in mask and hospital gown for his two-hour visit, he would try to feed Robby. Without any experience, he was petrified he would strangle him. Nothing else entered my mind except learning how to feed my baby. Sometimes I was encouraged and felt I was learning, and other times I was so afraid when it appeared more milk came back up than had gone down. At this rate, how could he gain weight and grow? These were the times I struggled with not picking up Robby and walking out the door, and I fantasized that I wouldn't even say goodbye. After all, it would be hours before we would be missed, if then.

I reached my breaking point one day when the on-duty doctor came by to see how I was doing. He asked me, "Are you taking hot sitz baths for your stitches?"

"I don't have time," I explained.

He interrupted me and said, "Well, if you don't care about yourself, there is nothing I can do."

That was the wrong thing to say to a stressed-out mom. I yelled at the doctor and told him I didn't need words but a little help from someone who knew more than I did. He promptly left. The accompanying nurse saw this new mom was very upset!

"I'll be right back," she told me.

In a few minutes, she returned with an older nurse I hadn't seen before. Introducing herself, she explained she had experience with feeding babies with a cleft palate and cleft lips. She was working on another ward, but she wanted to help me.

"You are doing a fine job," she said.

"Then why does more milk come up than goes down?" I interrupted.

She explained, "Looks are often deceiving. Your baby is making progress." After giving me a soothing and encouraging talk, she took Robby to look after him so I could get some rest.

"It isn't long before the end of my shift, and I will have to bring him back then. But this will give you a few hours of sleep," she said.

This was the only hands-on help I had for the five days I was there. I still remember my time in the hospital as one long nightmare. When Robby was sleeping or I was holding him, the overwhelming love and happiness I felt would come flooding in, and I would momentarily bask in the wonderful joy of being a mother. At a moment's notice, I'd have to concentrate once again on the best way to get nourishment down him.

Near the end of my stay, a doctor came in to discuss surgery for Robby. William Beaumont Military Hospital in El Paso, Texas, had a renowned surgeon on staff as a consultant every Friday. He had been a plastic surgeon during World War II and had plenty of experience. An appointment had already been made for us to take Robby to see him in a few weeks. The doctor kept reassuring me how capable the consulting doctor was. I felt better knowing there was a plan in place to help Robby.

After five long days, it was time to go home, and Bob came for us. Mother and Greta, my nine-year-old sister, were waiting at home, having ridden the train from Alabama to Texas. Going home and having company sounded so good! I had been lonely for too long. Being up night and day made it seem longer than it was. Your problems aren't as bad if you have someone to talk to. Mother cooked good food and did the chores. My job was to feed the baby and rest. I improved at balancing how much formula to give Robby before burping him. Huge volumes of milk still came up, but remembering the advice I'd been given that looks can be deceiving, I didn't panic.

Swallowing air can cause projectile vomiting, and from my shoulder he could throw it great distances. When it occurred in public, it got people's attention. Splattering on the floor, the milk was such an amount that it looked like it came from an adult. Everyone stared, worried, and thought he might be contagious. I carried lots of paper towels for quick cleanups. Heavy burp cloths were kept over my shoulder even when I wasn't giving him a bottle.

Robby thrived and gained weight. When we took him to see the plastic surgeon, the doctor outlined what surgeries Robby would need and the schedule for them. The first surgery to repair one side of the lip would be done when he was two months old. When he was four months, the other side would be repaired.

"How can he eat?" I asked. I quizzed the doctor about my concerns; he was patient and had answers. Deep down I was worried about surgery and a young baby being anesthetized.

The day for surgery arrived, and it was almost more than we could do to hand Robby over to strangers. I knew he needed the surgery. But he was so little. We were instructed where to wait and given an estimate of how long the surgery would take. To us, it seemed to take forever.

After surgery, the doctor came to the waiting room. "He is doing great and is in recovery. You can see him when he is moved to the baby ward."

Time moved slower and slower as we waited. Later a spokesman from the ward came to tell us, "Robby is awake and hungry—a very good sign. And as soon as he finishes eating, you can see him." With this encouraging news, our spirits soared that our two-month-old son was doing well.

When it was time, we rushed into the room. We could see Robby lying in a crib waving his arms. Getting closer, I could see what a miraculous job the surgeon had done. Tiny stitches in a perfect line were barely visible. He was so handsome. At that moment, I felt such hope to know this was possible, and that Robby was awake, looked happy, and was not in pain. He waved his arms wildly when he saw us. He had splints on both arms so he couldn't bend his elbows and reach the repaired lip. Relief washed over both of us, and we knew we could handle whatever was in store for us.

The next major life-altering event occurred approximately two years later, when I became pregnant for the second time. Applying my rule of "don't worry about what you can't control," I went on with my busy life. A few months into the pregnancy, I began to have the same dream over and over. It became more frequent as the weeks and months went by. In each dream, I was in the hospital, and the baby had arrived. I could see the entire baby except for its face and to determine its sex. A foggy blur prevented me from seeing the baby's face. It wasn't a disturbing image, although I couldn't see a single feature through the fog. I'd wake up at that moment and know my rule wasn't working. I was worried for this baby, knowing what could happen.

The dreams continued. In the last month of pregnancy, the dream occurred again, and this time I saw the whole body, the legs, arms, and the round head with two little ears. Then, for the first time, I saw the perfectly formed face. With a jolt, I woke up. The most glorious feeling swept over me. A miracle had occurred, and it had lifted my burden from me. I realized whatever happened, God was in control, and there wasn't any need for me to worry. Now I could concentrate on the coming birth and get mentally ready to go through childbirth again and become a mother of two.

Six weeks later I delivered a healthy baby girl weighing seven pounds, four ounces. She was two weeks late, and the labor was intense and fast. Bob came close to delivering this baby as we raced to the hospital. When we arrived at the entrance to the hospital, I leaped out of the car and told the workers bringing a wheelchair for me that the baby was coming. They shoved me into the chair and ran to a stretcher nearby, quickly transferred me, and I was in the prep room in record time.

The nurse was preparing to shave me when I again said, "This baby is coming now!" She looked shocked and then yelled for transporters to come back. The last thing I remember was being roughly transferred to the delivery table and having a mask put over my face.

I woke up with Bob sitting by my bed telling me we had a baby girl, and she was perfect. I kept drifting in and out of sleep. Each time I woke up, he had to give me the details all over again. We named her Diane. Every time I looked at her, there was the same perfect little face that had appeared in my last dream, reassuring me all was well.

CHAPTER 7

FACING ANOTHER LIFE-ALTERING DECISION

At sixty years old, I was facing another life-altering decision. I had to get my fear under control and meet this challenge of learning to ride my abused, starved, mad-at-the-world horse. I had worked with this angry horse for over a year, and I had observed little change in his mental problems. He was high maintenance, with a great need to know what I wanted from him. In all my years, I had seldom been this physically scared for my life.

Anyone who dealt with Marlon was convinced that he was mean and that he enjoyed reiterating that to them. Marlon didn't like strangers, and most everyone fit into that category. He put up with Bob most of the time because he often fed him, and Diane because she could ride well.

When his nervous energy was high, I was aware of it the minute I entered his stall. He couldn't be still, and he nipped at me continuously. On those days, I knew it was lunging time. If I was lunging in a round pen, I'd turn Marlon loose. I'd stand in the center of the ring holding a long lunge whip to give him directions and for protection if necessary. Walk is when the whip is at a ninety-degree angle to the ground. Lifting it halfway between 90 degrees and 180 degrees signals the horse to pick up the trot. Pointing the whip straight up means lope to a western horse and canter to an English horse. To ask the horse to change directions, the whip is pointed straight out, and you step closer to his running path. If the horse wants to challenge you, this is the most opportune time. Instead of reversing, he will speed up and race past you. When he is successful, it's difficult to get him to obey other commands.

If I was at a horse show, I hooked a long lead line to Marlon's halter and looked for a level place to lunge him. Often he'd start running around me as we walked toward an exercise area. He'd take off at a full gallop, tail held high, leaping through the air, kicking out, and making shrill sounds. He would run and run while I stood there holding one end of the lunge line. I wondered how long he could last at a full gallop. To a passerby, he'd look completely out of control. He was a big, powerful horse in the prime of his life. They knew that at any moment, he'd head for the hills when we went out in the open. I became the "lunging queen," because Marlon required so much of it.

On a day when my courage was up, Bob put the western saddle on Marlon, and I mounted my horse. As we walked around the pasture, I began to relax. Marlon was cooperating, so this might be easier than I imagined.

Being brave, I asked him to trot, but I bounced so much that after a few steps, I quickly said, "Walk."

"Sit down in the saddle and put your heels down. Now you are ready," I'd say, and again I'd ask him to trot. The bouncing started immediately, and I had to grab the saddle horn. Each time I asked him to trot, my bouncing made him speed up, and we'd have to go back to a walk. In that instance, it was obvious that I had an almost impossible task of learning to ride well enough to walk around the showgrounds and be somebody. Another major struggle was beginning.

I downplayed his behavior to the family, telling them all he wanted was kindness, love, and attention. No one believed me or excused his behavior. They continued to bitch about what "my horse" did to them, or they'd say they'd seen him misbehaving in some shocking way. Some of his antics I couldn't believe either, but I defended him. He was my horse, and he needed someone on his side.

Diane was Marlon's harshest critic. She complained the loudest about his behavior when she rode him. She felt Marlon was a vicious horse with no excuses for the things he did.

"It is up to us to figure out how to solve his problems," I told her, knowing that "us" really meant me. I kept repeating that we had to earn his trust and help him get his anger under control. At times I questioned if I'd live long enough to accomplish this.

Everything real or imagined spooked Marlon, making him leap forward and jump around. It was almost impossible for even an experienced rider to stay seated, and I was just a passenger in the saddle without a clue as to what needed to be done. Like with any sport, it takes a long time to get in shape and to learn to ride, and even longer when your horse doesn't cooperate. As time passed and Diane became busier with her young children and a part-time job, it became evident I'd have to get serious about riding Marlon. During the second year, desperate for help, I often drove up to Cathy's training barn for lessons and advice. Reluctantly, I'd bring Marlon back home. He had a new owner, a new place to live, and problems galore. I felt he needed to stay with me and have more time to adjust.

Marlon was taken to a few shows the first year after he had gained weight and was no longer anemic. At that time, we didn't realize to what extent his history affected him mentally. First we had to find a stall where he couldn't see a horse on either side of him, which was difficult. If he could see them, he constantly kicked the walls, ran in place, and never stopped. Not only could he hurt himself, he could get so hot and upset that he'd become sick. He had to prove to the other horses that he was number one in the pecking order. Once a horse was submissive to him, we could put him in a stall next door without a fight occurring. This was a serious problem. Showgrounds brought back terrible memories and could make him ill. Cathy and I tried every way we knew to keep him calm. Regardless of what we did, it was hard to tell if we were helping him. In desperation we sometimes lunged him several times in one day. Night after night, I replayed lunging sessions over and over in my dreams.

When I took my classy horse to shows, I'd be so busy handling his problems, I'd never have time to show him off. News spread quickly that my horse was out of control. Everyone watched when I led him into the arena to lunge him. We never caused an accident; he just looked so out of control. Adding to this picture was the tall, undisciplined horse running around me with such exuberance.

Trying to pick classes that Marlon tolerated best, Diane showed him a few times. In each class, he'd go the three gaits in the first direction perfectly, but when a reversal in direction was called for, he'd get anxious and speed up. She could manage him reasonably well until the canter was called for, and then he'd explode. Leaping into the air, kicking as high as he could with his hind legs, and squealing were his trademarks. Afterward he'd be calmer; the explosion released some of his tension. She tried to keep him away from the other contestants so he wouldn't interfere with their ride or hurt a horse or rider. They learned to watch out for Diane and Marlon. They kept a safe distance behind him or gave him a wide berth when passing. He felt threatened when horses rushed by him, and he'd again give his trademark performance. Often the other horses became upset, and a big ruckus occurred. Marlon could be expelled from the class for such behavior. This was very embarrassing for the rider. It meant they lacked the riding ability and the skill to keep their horse under control. Diane became good at keeping him from upsetting the other horses and riders, but she wasn't happy about having to do this.

Horses are judged on how well they move, but foremost they must have good manners. A good moving horse just floats around the ring at the walk, trot, and canter in both directions. They make the transitions from one gait to the next without a flaw. If a mistake is made, you probably wouldn't even notice. Marlon moved so well and looked so elegant, and Diane was such an excellent equestrian, that they'd often win or place if he was just halfway behaving. She was used to winning. She'd get so aggravated at Marlon when he misbehaved. Never missing an opportunity, she'd do her usual complaining after every class, even when I felt he had behaved decently.

"He is well trained, he knows how he is supposed to behave, and he does perfectly in warm-ups. Why can't he manage to behave in the show ring?" she'd ask me.

"He doesn't have the pressure on him in warm-ups," I'd tell her. "He is a smart horse, and he knows the difference. As I keep telling you, our job is to solve his problems for him. He doesn't know how." But Diane was tired of me making excuses for Marlon.

Going back to the barn area after watching Diane and Marlon in the show ring, I observed a lady and her horse walking up to a nearby stall. She was talking to her daughter when she quickly went behind her horse and snatched off its tail. I stood there not believing what I had seen. Despite the horse's naturally thin, stringy tail, she was holding a beautiful, long, full tail the color of the horse. I couldn't wait to tell Cathy about the artificial horse tail and was surprised when she told me that later Marlon would need one.

"Where in the world do you get one? What are they made of, and why does he need one?" These were a few of my questions.

The tails are made of real horse hair and come in different colors, lengths, and thicknesses. I was told a long, full tail helps to make a pretty picture as the horse flows around the ring at the different gaits. Right now, Marlon was dancing and jumping more than flowing. Preparing for that great day when I could imagine him flowing, I began pricing horse tails, and I ended up getting one for slightly more than a hundred dollars. I kept telling myself Marlon and Diane needed all the help they could get to look good in the show ring. I cannot repeat my retired-farmer dad's comment when he heard I had bought a horse tail for Marlon. He still hadn't gotten over us trading an automatic shift, air-conditioned van for a five-on-the-floor, stick-shift truck and putting a camper on the back to drive to Alaska.

When Cathy was present, she studied Marlon to discover what would work for this afflicted horse. It wasn't an encouraging situation for me learning to ride. I had to start taking one little step at a time, even though each one seemed insurmountable. As his routine was kept steady and he got plenty of good food, I imagined there was a tiny difference in his behavior. Knowing he needed it, I went to the extreme to keep his schedule the same day in and day out. If I was late with his feed, his normal behavior was to throw a tantrum when he saw me. Bob insisted I feed him on those days. Marlon's feed trough was in a back corner of his stall. As I walked back to put the grain in, he'd follow, threatening me every step across the large stall. Turning my back to him as I walked took a lot of nerve. I took chances on trusting him and learned which ones were the riskiest.

Toward the middle of the second year, Marlon was put in training with Cathy so she could try to get him under control. We went to the barn for riding lessons as often as my sore leg muscles could stand another day in the saddle.

It took me longer and longer to groom and saddle Marlon. Bob would have Dan groomed, saddled, and out into the arena before I seriously began. It became the joke that I arrived about the time everyone else was finished riding. There was a lot of truth to this; most days I was as scared as a person could be. Approaching Marlon, he warned me, and knowing a mistake could be made, I was always on guard. The time I spent grooming him, talking to him, and petting him helped settle my nerves so I could climb back in the saddle day after day. I hoped it helped him, but I knew it was impossible for him to totally relax while being stroked lovingly, even though he craved it.

To groom and saddle Marlon, we cross tied him by attaching a rope to each side of the halter and hooking the other ends to the wall or a pole. With his head movements limited, he still tried to bite me. He couldn't reach me, so he relied on his kick to do the job.

He was very particular about how you brushed him. You could never, ever make even one stroke that brushed the hair in the wrong direction. Touching his tummy was a definite no. If you dared to try, all of his ugly traits immediately popped out. It was amazing how quickly he could flatten his ears, do all his other moves, show his teeth, and then look around and snap at me. When angry or irritable enough, he'd lift his back right leg. If he felt he could get away with it, he'd do more than lift his leg, he'd kick you! He'd had lots of practice and was very fast.

Trying to do everything to his specifications, the next step was to sneak the saddle out and hope he wouldn't become violent. His worst nightmare was seeing the saddle and saddle pads. He didn't know who would be using them or what that person might do to him. Swinging his hindquarters from side to side when he saw them made it difficult to get them on his back. I dreaded this more than Marlon did, if that was possible. It took two people, and we usually had to throw the saddle on while he was moving and threatening us. Often we had to get the crop and let him see it, only giving him a mild swat if necessary. We didn't want to use it, knowing his past experiences and that he didn't take anything without fighting back. I scolded him and sometimes gave him an open-handed smack on his neck to let him know I was unhappy with him. His behavior petrified me, but I tried not to show it.

Tightening the saddle enough to keep it in place was the next challenge. After another big ruckus, I'd look at Marlon and question if he'd ever felt that anyone cared about him, or if I could survive learning to ride him. Such thoughts were counterproductive, though, and I worked at keeping them under control. Even on the worst days, I knew I couldn't give up on him.

After he was saddled, next came the bridle. He'd nip at me while I was getting the bridle ready and the halter off, but once I got past that, he'd open his mouth and take the bit. So thankful for this, I celebrated his good points.

Mounting the horse couldn't be delayed any longer. The first few times I climbed on Marlon, I did it from the ground and struggled to get my leg high enough to put my boot into the stirrup. Then I had to pull myself up to swing my other leg over the horse and sit down in the saddle.

"It wasn't pretty, but you made it," Bob said as he watched. Later I mounted from a bucket turned upside down, and this was much easier.

On riding days, the sun could be shining bright and warm, yet I could still find frightening things hiding in the shaded areas scattered about. Knowing that my horse and my safety depended on my being aware of my surroundings, this uptight rider also scouted the area for any loose objects lying around that could suddenly move. Looking for squirrels, cats, or dogs lurking in the area was my number-one priority. All dogs had to be put up until I finished riding. All gates had to be checked to be sure they were tightly closed, and if chains secured them, they must be taut, to prevent movement. I often fell off when a movement or sound caused Marlon to shy and turn suddenly. The family joked, and Cathy agreed (although she didn't say so), that I spit on my finger and held it up in the air to see if there was any wind that day. Sadly, this was true. If there was enough wind to rattle or blow things around, it was almost impossible for me to ride.

"She thinks she can't ride if a worm is passing gas because the noise would be enough to frighten Marlon," Bob commented.

Dan was shorter than Marlon, and Bob towered over me, but the thought never entered my mind to change horses. Marlon was mine. Falling off did make me realize how tall he really was, though. It was a long way to the ground, and I was aware of every second of a fall. I'd start bouncing, and with each bounce, I'd slowly lean to my right and slip down Marlon's side. How much it hurt depended on the hardness of the ground. In heavy sand or sawdust, it wasn't too bad; on grass or packed dirt, it was painful. The physical pain was easier to handle than the games my mind played on me. "Thank you, Lord, for looking after this foolish, sixty-plus-year-old woman" was my constant prayer. To me, nothing else could explain why I didn't break a bone in falling onto the hard ground from such a height.

I had admired lush, freshly mowed pastures, groomed to perfection, with fences in every direction. Now I lived in this very setting, with pastures surrounding my house and visible from every window. But my enjoyment was tempered by my neurotic horse. Looking at the view, walking to the barn to do chores, or weed-eating those miles of fence lines, I could temporarily forget how volatile my situation was with Marlon. Arriving at the plain, modest barn that we had built, or thinking about riding, was all it took to remind me and bring me back to reality. Usually I was not one to worry about what might happen, yet when riding Marlon, it was hard to convince myself it would go well and I'd be safe. Most of all, I wanted to enjoy riding him, but I couldn't see this happening anytime soon. Bob was concerned about my situation, although he never asked me to give up or not ride. Occasionally on windy days, he'd ask if I intended to ride. I never said no; I was afraid it might end my challenge forever. Sometimes he made the decision for us, saying, "We can wait for a more reasonable day to ride." I never objected.

Bob knew my background of growing up on a farm without electricity and how hard I had worked to survive and go away to college. He also knew I never gave up until every option had been explored. He'd remind me of this on the bad days.

While the horses were away and being cared for, we had a chance to travel and go see our children, who lived out of state. Several years after Marlon came to live with me, Diane and her family moved to Maryland. It was another heart-wrenching occurrence in my life. Not only did I lose my knowledgeable horse-helper daughter but also my granddaughters, Jessica and Rebecca. They had lived all their lives across the pastures and fields from us. I had to work night and day to remind myself it wasn't the end of the world, although it felt like it. I was so distraught about the granddaughters moving away, I composed a sorrowful poem and sent it to them. When Diane read the poem to the girls, they all cried.

She called me and said, "Don't send any more sad poems. We are sad enough!"

At least everyone was healthy. Mike had a job; Rob was a pilot, and we could fly standby to go see them. Like everything in life, it took time, but we all adjusted.

When we visited them, they probably got tired of hearing about our current Marlon and Dan problems, but they graciously tolerated it. My precarious situation with Marlon dominated the conversation. With it constantly on my mind, I found being able to talk about it helpful. It didn't seem as bad retelling it in a pleasant and fun atmosphere.

We were laughing and telling funny stories about each other when I reminded Rebecca about the day we were at the barn, and she'd kept staring at me. When I'd catch her eye, I'd say, "What?" She kept saying it was nothing. Then she finally said, "You don't have as many cracks in your face as my other grandmother."

She didn't admit remembering this, but she laughed. We agreed laughter helps you handle problems.

Occasionally, when I rode under Cathy's supervision and things had gone well that day, I'd feel a little more encouraged. But it was such hard work. Most riders used metal spurs on the back of their boots to help cue the horse. If used properly, spurs help the horse understand what you are asking him to do. It is difficult for the horse when you are bouncing around, learning to ride, and building up your leg muscles. For years, I could never wear a pair of spurs to cue Marlon; it brought back terrible memories for him. Unscrupulous trainers often use large, jagged metal spurs, and Marlon had similar experiences in his past. If he felt metal, he'd go crazy. I discovered this when trying quarter-inch metal spurs, trying to get some relief for my sore legs. It was a terrible mistake. He reared up on his back legs, making me feel we were falling sideways, and then he brought his front feet back to earth and immediately kicked his hind legs out. With the rocking motion and the other moves he threw in, it was terrifying. I was riding in a western saddle and only escaped disaster by grabbing the saddle horn and hanging on with all my strength.

"How much longer will it take for us to get in sync and better understand each other?" I often asked Cathy.

"You have to build muscle memory as well as develop the strength for riding," Cathy told me. "Then there is the issue of earning his trust, but it is slowly coming."

When Marlon was having a calm day and could think straight, he quickly learned he could take advantage of me. He did what he wanted instead of what he was asked to do. This was a real concern. I'd been told repeatedly that I must be in charge, or he'd never respect my authority.

"Smack him so he'll know who is boss," Cathy would sometimes say.

It was a long time before I could give him an open-handed little smack while on his back. I was too afraid, if truth be told. I made excuses to myself that the timing wasn't right, or that it was after the fact and he wouldn't understand if he were smacked now.

Thinking back, my worst fear was not falling off but being in the saddle on a horse totally out of control. I was doing everything I knew to do, and it wasn't helping one bit. It felt like I was on a giant roller coaster. It was moving up the incline at a controlled speed. Once I made it to the top, I'd blast off, and anything could happen. Never knowing the moment the top would be reached caused great anxiety for me. It hampered my ability to properly apply what little knowledge I possessed. It took a long time for me to get over this and to stop feeling that the earth was whirling around under me anytime I was on my horse.

Flight is a horse's natural response to being scared. Marlon was always on alert, with his ears up listening and his eyes scanning the landscape. He was light on his feet and stayed ready to run. One of the scariest things for him was birds or planes flying overhead and casting shadows that constantly moved. He couldn't figure out what it was or where it was. At one time, our riding arena was in the practice-flight path of military fighter jets. In my early days of riding, when we heard a plane coming, I'd get off. When I could ride it out, I was making progress. Another time, we rode near a highway, where the big trucks deliberately made loud backfiring noises just to see how the horses would react. One of my falls came about as a result. Cathy accused me of not thinking ahead and being ready when I heard the truck coming. Either that was true, or I was too scared to prepare for the inevitable.

One day, as I was riding and struggling to stay in the saddle, I began my usual pattern of bouncing, leaning to the right, and falling down Marlon's side. From the moment I knew hitting the ground was inevitable, my fear conjured frightening scenarios from the past.

In 1986, on a snowy Super Bowl Sunday, I arrived at the base of Chapman Mountain, east of Huntsville, Alabama, around nine in the evening. Cars and trucks were covering the road as far as the eye could see. They were in the road, on both sides of the road, facing forward, backward, and sideways. Heavy snow was falling, and visibility was getting worse by the minute. At that moment, I knew I could not get over the mountain to go home. I immediately pulled onto a side road where my car would be safe. There were

The Big Bay Horse Doesn't Live Here Anymore

houses lining the road, and one of them had its porch light on. I walked up and rang the doorbell. A young man who looked to be in his twenties opened the door. I told him my problem and asked if I could use his phone. He led me back to the kitchen, where there was a wall phone behind the kitchen table.

I called home, and Diane answered. She and three-month-old Jessica were spending the night with us. Diane repeated my story to Bob. He ran out to the woodpile and started loading the back of our pickup truck with firewood. He hoped this would give him traction to get over the mountain to pick me up.

Diane knew a girl who lived in the area I was calling from, and she said, "I will call her and see if you can spend the night. Call back in a few minutes."

I hung up and turned around to tell the young man who was sitting at the kitchen table that I wanted to wait and call again. Before I could say anything, I saw a rifle and a pistol lying on the kitchen table in front of him, and he was checking them out! I was petrified. Thinking fast, I asked him about his neighbors.

"An older couple live over there, and they have grandchildren visiting all the time," he said as he pointed to the east side of his house.

He had said the magic word when he mentioned grandchildren. I thanked the young man, immediately left, and went next door. I wasn't dressed for snowy weather and still had my name tag on from a church meeting that I had attended earlier in the evening. I rang the doorbell, and a ten-year-old boy, followed by an older gentleman, opened the door. A lady joined them as I was telling my sad tale.

We talked a few minutes, and then she said, "Do you want to use the phone?"

"Yes, but I want to sit on your sofa all night!"

They started laughing and said, "We can do better than that; we can give you a bed and some clothes to sleep in." I was so relieved and thankful. I kept muttering my thanks over and over, with tears streaming down my face.

I was thrown from a horse only one time, but I had many falls, and each flashback was different, some more traumatic than others. As I hit the ground, I always wondered, "Will I ever ride again? Will I be hurt so bad that I can't care for my horse or live my life as I know it?" Getting up and walking around let me know that once again, I had survived unscathed.

CHAPTER 8
RIDING LESSON

Cathy was giving me, Bob, and several other clients a riding lesson in an outdoor ring. It had been raining and the ground was slushy, although the footing was good in the sandy soil. We were all loping in a large circle, and just as Marlon and I rounded a corner, one of his back feet slipped. This caused him to go straight. I continued leaning around the curve and fell off into the muddy slush. Wanting to help your horse around a curve is natural, but it hinders the horse instead of helping. A rider must always sit up straight regardless of what the horse is doing. Another important lesson was learned that day. My fellow riders dried me off as best they could. Muddy as I was, I continued my riding lesson.

Each time I fell off Marlon, I got up, moved around, and checked to see if my body parts worked as well as they did before the fall. I wasn't young anymore, and it took time to evaluate whether a little pain meant anything. As I walked around to recover from a fall, a scene I had witnessed long ago in Anchorage, Alaska, often floated across my mind. One snowy winter day, I had driven Diane to the barn to see about her horse. An army sergeant who had a horse at the military stables came to the barn, put the bridle on his horse, and led it into the outdoor arena. There were several inches of snow on the ground, and snow was still falling. He stopped in the middle of the ring, threw the reins over the horse's head, put both hands on the horse's back, and gave a big leap. Over the horse he went, doing a complete flip. He landed on his head, then toppled over onto his side. Lying there only a second, he jumped up and looked in all directions to see if anyone had seen him. He couldn't see me looking out the window and laughing. Everyone in the tack room tried to figure out what was so funny. By the time I could talk, the attention had moved on to something else. The sergeant's secret was safe from the people who knew him, but not from my family and horse friends. We shared many laughs because we could imagine this happening to us. I couldn't resist recreating the scene for Cathy after one of my falls.

At the end of a riding lesson, I was discussing my Marlon problems with Cathy. "Most of your clients fall into the geriatric group. Don't you get tired of telling us the same thing over and over?" I asked. "Wouldn't you rather have younger and quicker-to-learn students?"

"I like working with older people," she said, surprising me. "When they learn something new, they are so appreciative, excited, and thankful! The younger students want to learn it all in a day. They are often out of sorts if they don't."

Her words were encouraging, and I was happy to have survived another riding lesson. Looking at Marlon standing relaxed now that his work was finished, I was momentarily on top of the world. Such mountaintop moments were fleeting at this stage of my riding.

One day it dawned on me that I couldn't ride English, and English was Marlon's specialty. Changing from the western saddle would be difficult for me, but most of his winnings had been in the English riding classes. He liked doing what he knew how to do; thus he tolerated the English saddle much better than the heavy western saddle. Most trainers use the western saddle, and past treatment by some trainers was still fresh in his mind. I knew it was important to switch to his preferred way before going any further in my training.

Nodding her head and thinking about it as I presented my case, Cathy finally said, "OK," even though she wasn't keen on my idea. She knew I wasn't experienced enough at riding to contemplate such a transition. This started me on the long road to get in shape to do English posting, rising up and back down in rhythm with the horse as he trotted. When riding western, it is natural for a new rider to sit back in the saddle and let their legs go forward. But in that position, it was impossible for me to raise my body up and hold it for those few seconds as Marlon took a trotting step. The object is to be one with the horse as he moves. I worked daily to get my legs back far enough and have the strength to rise up as my horse trotted. After months of practice, Cathy would inform me that it was time to raise my stirrups another notch. This was good and bad. It meant I had progressed, but it felt like starting over in the conditioning of my legs.

Marlon could feel when I was off balance, and he'd speed up, thinking, "What is she doing, and what is she asking me to do?" Later he'd realize what was happening, and he'd slow down so I could recover, if I was still on his back.

"Poor Marlon, look what he has to endure" was Diane's comment as she witnessed my struggles. These were the times she had sympathy for him.

Cathy told me that she was surprised I stuck with the English riding. Most of her students decided to change over to English riding at some point, but after a few tries, they usually went back to western. In the beginning, it takes longer to get in shape, and the saddle is smaller. Knowing my crazy horse preferred this saddle motivated me to struggle onward.

CHAPTER 9
FIRST HORSE CLINIC

Bob had been riding for a few years when he saw an advertisement about a famous judge and horse trainer teaching a nonviolent way to achieve unity with your horse. The ad stated, "Trust between you and your horse must be established before it will want to please you." Bob decided he was far enough along in his riding for such a clinic. He was always charging forward whether he and Dan were ready or not. I decided I would attend the clinic with Bob. Spending a week with your horse and other riders sounded like fun. Bob and Cathy thought I'd go as an observer, but that never entered my mind. Cathy wasn't enthusiastic about me taking Marlon. She knew I didn't have a clue about what was required of a rider at one of these clinics. Lack of knowledge didn't dampen my enthusiasm, though. Adding to our lack of riding experience, Bob started with a young horse, and mine was an angry, disturbed one who had been abused.

The article stated the clinic would start on Monday morning, end on Friday afternoon, and be held in Gainesville, Georgia. Bob gave me time to change my mind, but when he realized that I was serious, he registered us for the clinic. Refusing to clutter my mind by questioning what the five days of riding would entail, I started collecting food for the horses, buckets for them to eat and drink from, and all tack needed for a week of horseback riding, and I packed them into the horse trailer. The advertisement claimed all clinics were held in an indoor arena, so we thought the cold, snowy month of January wouldn't be a big factor. Getting ready to go didn't leave me time to worry about my decision. If Bob worried, I couldn't tell.

Sunday afternoon we left for northern Georgia and arrived to find snow on the ground, very high winds, and only a covered arena. We were shocked that the arena was not enclosed as advertised. We unloaded our horses and found stalls for them. Only a few horses were in the barn. Walking from the barn to the arena, we noticed that big metal garbage cans lined with plastic bags surrounded the arena. The wind was playing a tune on these cans, with the bags flapping and slapping the sides, starting and stopping at random. The noise was deafening, and I knew Marlon would be petrified. At that moment, I had an inkling of what the days ahead were going to be like.

Sam, who was holding the clinic, arrived. We introduced ourselves. Although he was famous in the horse world, he was friendly and immediately put us at ease. As we chatted and walked with him through the barn, we heard a loud noise. "What in the world was that?" I said.

"It is that big angry horse over there," replied Sam. Of course, he was talking about Marlon. It was one more bit of evidence that the week wouldn't go well for Marlon and me. From the look on Sam's face, I think he realized anything could happen at this clinic with the bad weather and an open arena.

A short meeting was held that evening for the ten brave souls present to meet each other. Sam apologized for the arena not being enclosed and explained that a local conflict was to blame. He outlined the schedule for the week and gave us a pep talk. The mornings would consist of ground work. He emphasized the importance of learning ground work to teach us important basics before we moved forward with riding. After the lunch break, we would saddle our horses and be ready for riding instructions.

Monday morning was colder than the day before, and the wind was extremely high. Marlon and Dan wore heavy winter blankets; we had to remove them for the clinic. Doing ground work, they weren't moving around enough to keep warm. The first morning, I discovered my horse didn't care for ground work. He told me about it every second. He'd do each exercise once, and when asked to repeat it so we could master the technique, he'd stubbornly refuse. Throwing his body around in such a way that I couldn't tell what he was doing, or threatening to bite my hand holding the lead rope, were his tactics. If his mission was to unnerve me, it worked.

Why didn't we leave one of the lighter blankets on the horses for the ground work? We were too consumed with our problems to think of it, and Sam never suggested it. Bob fared much better because Dan cooperated even if he was cold. I was so discouraged by noontime.

We went to lunch, commiserated with each other, and traded hard-luck stories. One lady told us when she went to the pasture to get her horse for the clinic, she couldn't catch it. Not wanting to lose her money, she caught a different horse and brought it. Unfortunately, the horse hadn't been saddled for over a year and was out of practice. Another lady had brought a racking horse, which had an extra beat in its gaits. This made it strike the ground harder and have an entirely different sound to its movements. She wanted to teach it to drop the extra beat. Each rider and their horse had a story all their own. My story was to survive the week.

After lunch, we saddled up and waited in the arena for Sam to join us. He instructed us on how to properly mount our horses for the afternoon of riding. After mounting, we started walking around the arena. Immediately, Marlon picked up on the odd sound of the racking horse, and my problems escalated. It scared all the horses at first, but Marlon thought an alien had come for him. He never completely adjusted to the sound all week. I moved as far away from the horse as possible. Marlon was causing such a ruckus, no one wanted us within a mile of them. We didn't have a place to go; even Bob and Dan didn't want us near them.

"Move away, you're upsetting Dan!" Bob said.

I sure understood how he felt, but it hurt my feelings for him to say that when I was so desperate. If it was too much for my own husband, then my problems were hopeless.

All the horses were skittish, and the riders didn't want anything to further upset them. Sam's solution was to send Marlon and me to a small enclosed area at one end of the arena for some circular work. We went so often Marlon tired of it and refused to enter the gate. I was at my wit's end by the time the afternoon was over. Everyone kept encouraging me at the evening review. I was the problem child, and I didn't like it one bit.

Back in our hotel room later that evening, Rob and Diane called to see how we were faring. From the beginning, they'd been concerned about me and Marlon participating in the clinic. My tale of woe was a sad one, and I was as low as anyone could be concerning my problems with Marlon. Thinking about four more days of the same, I prayed for courage and for the wind to die down.

The next couple of days were more of the same. With all ten horses and riders in the arena, and all going at different speeds, there wasn't any place for Marlon and me to stay away from them. My number-one priority was to stay as far away from the racking horse as possible. It was so frustrating and scary for me. Marlon was unhappy, and he used every method he knew to tell me how he felt. I got his message. There wasn't anything I could do to help our situation. All I could say at the end of each day was, "I gave it my all! I survived another day, and I didn't fall off."

"Do you have a western saddle?" Sam asked me on Thursday afternoon when we were ready to saddle our horses.

"Yes, I have one in the trailer." He suggested I use it. Why hadn't I thought of this?

The wind had died down, the sun was out, and it wasn't quite as cold. The snow was mostly gone. We moved to an outdoor riding ring without the garbage cans. Compared to the other days, the afternoon was a breeze; I could relax a little and not be so scared. It helped that Marlon's attitude had improved. I felt safer in the western saddle, and we had better riding conditions. My fellow riders gave me lots of encouragement at the evening gathering, remembering how much I had improved.

Friday morning, we had a short review on the ground work we had practiced and went over new things we had learned while riding. Then it was exam time for our ground work. As we followed Sam's instructions, he graded us. Next we saddled our horses and were tested on our riding ability and how much we had improved. After a late lunch, it was time for the final critique. Sam went over the exam and discussed our positives and negatives with each of us.

I passed is all I can say. Everyone laughed and discussed the many problems I'd endured all week. They gave me lots of compliments for surviving and completing the clinic. I left feeling I had learned a lot, but I knew I never wanted to go through such a terrifying ordeal again. Becoming a better rider and helping Marlon calm down were my goals, and I hoped for this opportunity again.

At my second Resistance Free Training clinic a few years later in Florida horse country, a funny incident happened.

"Each person will ride individually in a round pen without a bridle to demonstrate how your horse will heed your leg cues," Sam told us.

I refused. Marlon was still too unpredictable. Sam decided one of his trainees who was helping him run the clinic would lead Marlon with me sitting in the saddle. Since this was to build my confidence, I agreed. The trainee was out in front of Marlon at the end of his lead rope. We had walked one time around the ring when Marlon evaluated the situation and decided she wasn't leading him properly. She should be walking by his head. Suddenly he trotted up to her hand holding the lead rope and bit it. A big commotion ensued. The trainee was indignant that he was so undisciplined he would bite her, plus it hurt, and her hand was bleeding. For the remainder of the clinic, she was unhappy with me and Marlon.

The later clinics were great fun, and we learned a lot, both things to do and things to never do. Each time we attended a clinic, Marlon and I were further along in our training. After spending an uninterrupted week together, I understood him much better. If I could predict his behavior, then our trust in each other would improve.

Horse trainers who practiced Resistance Free Training were often referred to as horse whisperers. After I had attended a couple of these clinics, I was following a pickup truck and noticed a bumper sticker with a picture of a horse and writing underneath. I wasn't near enough to read it. I pulled closer and read, "I whispered, but my horse didn't listen!" Laughing, I thought, "That sums up me and Marlon."

CHAPTER 10

MISSISSIPPI HORSE SHOW

I took Marlon to a horse show in Mississippi, along with Diane, Bob, and Dan. We went with Cathy and some of her other clients. Diane was planning to show Marlon in a few classes. There were three shows over Friday, Saturday, and Sunday. We arrived a day early to find our stalls and to make sure they met all the required specifications for Marlon. We unloaded our trailer, got the horses out, and introduced them to the barn area, the outdoor riding rings, and the covered arena. The show would be held in the indoor arena. This was the first time I felt confident enough to get on my horse and ride around the showgrounds. It was a big deal for me. Cathy and my family strongly encouraged me, as this was a giant step forward.

Trailers with horses were constantly arriving. Some of the people were already out riding or exercising their horses. The showgrounds covered a large area and were truly beautiful. Flowering shrubs and blooming plants were located along the outer edges of the fence lines, safe from the working horses. New places were always scary for Marlon and me. I hoped to settle my nerves by concentrating on the beauty that surrounded me. I lunged Marlon in one of the large outdoor rings to let him see the area, kick up his heels, and hopefully get rid of his pent-up energy and nervousness. After exercising him, I put the western saddle on in case I needed the saddle horn to hang onto, and then I mounted my horse and started my adventure.

It was a bright spring day without a cloud in the sky, warm but not hot—a perfect day to ride my special horse. We walked out to the warm-up area and joined the other riders. Marlon looked all around and didn't see anything he couldn't handle with ease. He was calm and happy. My confidence level increased.

After we walked around for a while, a brilliant idea popped into my head. "I believe I'll take Marlon into the covered arena where Bob, Cathy, and the other horse show contestants are practicing," I thought. "Hey, I'll show them how well I'm doing."

As we approached the big double doors leading into the arena, a horse and rider blasted through them, making a loud noise and flying past us. This scared both of us and really upset Marlon. I tried to soldier on. I talked soothingly to him and told him it wasn't any big deal, as the horse and rider were long gone. Walking closer to the entrance, I encouraged him to walk through those doors into the arena. He planted his front feet and wouldn't go forward. Each time I encouraged him to move toward the entrance, he'd quickly turn around and try to go in the opposite direction. We circled around and around, and then I asked him again to walk through those doors. His refusal was more adamant each time. With my limited experience in riding and dealing with a problem horse, I tried everything to convince him he must walk into the arena. Instead he ended up convincing me that today, he wouldn't walk through those doors.

"The decision has been made, and we can just as easily ride outside." That was the idea flashing through my thoughts in big neon lights as we turned around to leave.

This unnerved me, as I'd failed to get the message across to Marlon about who was in charge. Approaching the warm-up area, I struggled to regroup my thoughts and decide how to proceed after my dismal failure. The warm-up area was empty except for one horse and rider. After walking around to calm our nerves, Marlon and I suddenly heard an uproar going on behind us. Glancing over my shoulder, I saw the rider whipping his horse. The horse was fighting back, rearing up on his hind legs, jumping around, and trying to get away. Marlon sensed what was happening and started fretting and twitching. His skin was crawling underneath me, and it was evident I had to get him out of there fast. To exit, we had to walk closer to the horse and rider. By the time we got back to the barn, Marlon was very upset. I was feeling so low, knowing I had let him down.

Unsaddling Marlon and taking him outside to spray him off, I noticed he'd broken out in hives. His nerves and the flashbacks had become too much for him. The remainder of the afternoon was spent trying to calm Marlon. He was a pitiful sight, constantly twitching and alternating between letting me love on him and throwing his head up in the air, looking for the bad guy. His eyes looked so sad, and you could see such pain in them. As he begged me for relief, I felt so helpless that I couldn't ease the terror in his mind. As the afternoon dragged by, he'd put his head up against me and hold it there for a minute or two. In that moment, he was thanking me for being the truest friend he had. By feeding time, his pain was unbearable, and the vet was called. The flashbacks from his years of mistreatment were too fresh in his mind, and he became colicky from the stress. I kept an all-night vigil, making sure he was progressing and didn't lie down. After treatment, the vet wanted him to stand until he was feeling better. The pain of colic is excruciating, and horses want to lie down and thrash around when they are afflicted with this condition. A twisted intestine can be the result, making surgery necessary.

Marlon was feeling better the next morning and could eat a small portion of his grain. We gave him plenty of grass hay. He spent most of the day eating and taking little naps standing up. His nerves still wouldn't let him relax enough to lie down. It took several days for him to recover.

Back home I had to start working on me so I could continue riding. Recovering from having your confidence shaken takes time, and there wasn't time. Good experiences between Marlon and me were needed in the weeks ahead to help me go forward. To continue as if nothing had happened was tough. I hoped my bravado covered what I was feeling inside. In time, I progressed emotionally and improved in my riding ability. It was like two steps forward, then a step and a half backward, and some days several steps in reverse.

Before going to the next horse show, Bob suggested we should look for a small motor scooter to help with all the walking required at horse shows. The barns are usually a long distance from the arena, and you can literally be too tired to ride by the time your class is called. This sounded great to me. There was a large motorcycle shop not far from us, and we went to look. A new one was rather expensive, but they had used ones. As luck would have it, there was a hot-pink scooter with only fifteen miles on it. We looked, debated, and finally bought it. Bob received all kinds of teasing as he raced around the showgrounds on his little hot-pink motor scooter. He could take all the joking when he thought about the price he paid and what a lifesaver it was. The granddaughters loved for Papa to give them a ride, and later they could ride it alone.

When dealing with Marlon and learning to ride, I wanted to tell family and fellow riders my unbelievable horse troubles. No one understood what I was going through. Dealing with Marlon's mental state twenty-four hours a day, and my constant fear and worry, were difficult for others to grasp. They thought I was just a little old lady who was exaggerating; however, they knew I believed it. Everyone was aware of Marlon's many real and imaginary problems, but they didn't have a clue how severe they were.

Cathy was the only one who understood. When something happened so baffling it was beyond my vocabulary to explain, I'd ask her, "Why did he do that? What does it mean? What can I do about it?" Being a talented trainer, and having been around horses all her life, she had a practical, common-sense approach to handling horses. But some of his antics stumped even her.

I often dreamed I was riding Marlon when one of our disasters occurred. I'd try to hang on, get him under control, and reassure him that we were partners. I'd wake up in the midst of trying to solve our problems, deeply disturbed and asking myself, "Will he ever improve?" I kept hoping a solution would come to me in one of these dreams.

Some dreams were pleasant. These dreams occurred when Marlon and I had worked hard and he'd cooperated with me. They were such sweet dreams that I could have leaped up and danced. A little success gave me needed encouragement and glimmers of hope and joy.

My attitude at any given time determined my response to Marlon's behavior. From the beginning, I felt he was doing all he knew to do, and he needed my help. This was a new chapter in his life, and a horse never forgets bad things from its past. Under the right circumstances, the past can come back in a flash. If he could forgive the people from his past and start trusting again, wonderful things could and would happen. I hoped to lessen the recurrences with lots of love, patience, and hard work.

The toughest thing was my fear, which could turn to sheer terror when I was riding and something scared or upset Marlon. The times I managed to stay on his back and not fall off gave me encouragement to keep riding. Bob gave a running commentary on my ability to scratch and claw to stay on his back. "My gosh, you made it! I can say you are getting better day by day, but it still ain't pretty," he would say.

It was hard for Bob to see me fall, or struggle not to fall, almost every day. He was trying to add a little humor and to inspire me to keep riding. My balance improved, and that helped. My confidence level was determined day by day, without rhyme or reason. On bad days, it was so tempting to find an excuse not to ride. But I couldn't give up; excuses are habit forming.

Having a show horse at an older age and learning to ride was tough. You had to be passionate about it, or you wouldn't continue with such a hobby. The work alone was overwhelming. Riding conditions could often be described as awful. We rode when it was so hot, I thought I would pass out from the heat, and I could easily drop five pounds in a day. Riding in ice storms, with sheets of ice sliding off trees and buildings, terrified me and Marlon. If Marlon was scared, then I was scared. The gale-force winds were the worst for really upsetting him. Because the winds were unpredictable, you never knew when they'd start, but you knew they were coming.

Our motto was similar to the mailman's: "Neither snow nor heat nor rain nor sleet nor wind will keep us from riding our horses." Repeating this gave us a good laugh on those bad-weather days.

It was always fun to ride with Cathy's clients. She had a group of regulars, while other clients came and went. Arriving at the barn, new riders would be ready to join our group. They'd be thrilled with their new horse, the new tack, and learning to ride or starting to ride once again. But so many quit before they gave it a chance. Often it wasn't what they thought it would be; too much work was involved, or they decided their horse wasn't a winner in the show ring. Everyone wants to win. Bob and I were with Cathy for nineteen years and saw many riders come and go. We became good at deciding who would stay or who would leave in a hurry. Sometimes Cathy was overjoyed with a new client because they had the right attitude. They were willing to work hard, and their expectations were reasonable. Over the years, I'd ask Cathy if she thought a new client would make it. She had years of experience, was excellent at evaluating new customers, and was usually correct.

A pretty lady in her forties arrived one day with a gorgeous new horse that was a winner in the show ring. She had the best tack money could buy. She showed great excitement about starting to ride again after riding when she was young. Several riding lessons later, I asked Cathy the question I always asked. "Do you think she will make it?"

"Oh yes, she has what it takes!" Then she went on to list all her qualities.

Only a few months later, though, the horse was gone from the barn. The lady had decided she didn't want to be a horse owner. Her beautiful horse was for sale. Cathy and I couldn't believe it; she seemed to have the desire needed. But the hard work required is so much more than people imagine. It takes time to strengthen your body for riding, and the results are slow in coming. Bob and I understood this, being older riders.

CHAPTER 11
ENTERING THE BEGINNER CLASSES

My horse showing progressed at a slow speed. I started entering the novice Hunter Under Saddle class, strictly for beginners. As soon as possible, a contestant moved on to the amateur class. It took us older riders with problem horses much longer. While still a beginner, I went to Perry, Georgia, to a horse show. It was a six-hour drive, so we left home early in the morning. As usual, we needed plenty of time to rest, set up the trailer, let the horses adjust to the new environment, and practice riding.

The show was held in a state park with campgrounds and an RV area. The first evening, Cathy told all her clients to saddle up and join her in the outdoor covered arena, which had plenty of bright lights. This added to my nervousness; I had never ridden in the evening. When I looked at the well-lit ring, I couldn't help wondering what Marlon would see in the dark beyond it. I reluctantly obeyed.

I joined the other riders and started walking Marlon around the ring. I talked to him in my calmest, sweetest voice, trying to relax both of us. To my great surprise, Marlon was rather relaxed and listened to me. We started trotting, and thoughts flew through my head that we could do this. After settling down a bit, I noticed a crowd of people standing around the outside of the ring. More and more people kept coming. When we changed directions and slowed down, some of the people spoke to me as I passed by. I just smiled and continued listening to Cathy's instructions.

When the practice session was over and we were walking out of the ring, a group of senior citizens met me and wanted to talk. They were so complimentary and encouraging to this fellow senior citizen. They were attending a week-long RV convention in the park. I had a ringside cheering squad for the entire weekend

Later we decided to go to a horse show in Montgomery, Alabama. Rob, who lived out of state, was worried about his mom with her wild horse. Since I had suffered several falls, he was concerned about me. When he heard we were going to Montgomery, he sent word that he and his family were coming.

I was so surprised, as he always said, "I've been to enough horse shows to last a lifetime." Having a horse when he was young convinced Rob that they were too confining. He preferred other activities. Also, he'd spent years going with us to watch Diane compete in shows before he became old enough to stay home alone. I was happy he and his family were coming but puzzled as to why. Later Rob's wife, Karen, explained that Rob wanted to see for himself how I was doing.

Competing in a class was new to me, and I wanted to do a good job with Rob and family present. Julianne, my oldest granddaughter, and her younger sister, Jeannene, would be watching their cowgirl grandmother compete in a horse show on her out-of-control horse. The pressure was on for me to rise to the occasion and set a good example for them. I had spent much time explaining Marlon's behavior to them as they stood a safe distance away, watching and listening. The belief that impossible things can be accomplished if you work hard and give it your all was one of my favorite things to tell my sweet granddaughters.

They talked and asked more questions than I could answer as I groomed Marlon and saddled him for my class. Then it was dress-up time. The girls were excited when I appeared in my English show clothes. I led Marlon from the barn up to the indoor arena where all the other contestants were waiting. Waiting is always the worst part. You go from being calm to being in a panic, and then you go back and forth until it is time to enter the arena at a trot. You hope you are in your calmest mode when your class is called. Entering at a trot scatters the contestants around the ring so they can be seen by the judge. This was a large beginner class.

Suddenly it was time, and the entire family rushed into the arena and took seats in the bleachers. Entering the arena and picking up a fast trot, I could see smiling, expectant faces in my peripheral vision as I flew by. "Walk," the announcer said when the contestants were spread out around the ring. We were now being judged. The horse must walk with a steady gait and have perfect body carriage. We walked for a while, and I managed to stay away from the other riders, or they avoided me. I began to get nervous. The trot was coming up, and Marlon was inclined to go faster than he should. You must move out at a fast, controlled, steady trot, but not like you are running a race.

When the announcer called for the trot, I kept Marlon at a reasonable speed, even though we passed everyone.

Nervous or not, I loved going at a fast clip; it was so exciting, even if it wasn't what I was supposed to do. The announcer called for the contestants to walk. This gives the riders time to settle the horses and prepare them for the canter. I managed to prevent Marlon from giving a wild leap into the canter when it was called for. He was too unpredictable for me to let him canter. Instead he dashed off at a trot so fast, I could hardly rise up and down quickly enough to keep up with him. The same scenario continued for the last half of the class. Finally we lined up in the center of the ring for the judge to check our backing abilities, and the class was over.

I was pleased with another learning class where there hadn't been an incident of any kind, and I was more determined than ever to get better. Afterward, we all stood around talking and laughing about my ride. Cathy was glad I had read the situation correctly and hadn't let Marlon canter.

"It wouldn't have been pretty," she said.

She had worried that I was not experienced enough to even enter the class. The girls were bright eyed and happy because they had seen their grandmother compete in a horse show. I ran into a friend, who congratulated me on entering the class.

"There was one rider who never cantered in either direction," he said in disbelief.

"That was me. I couldn't take a chance and let Marlon canter."

He was taken aback after hearing that. Marlon's horse show reputation, as well as mine, was enhanced that day!

Later, in addition to my Hunter Under Saddle class, I added a Western Showmanship class. This isn't a riding class. It's judged on a horse's appearance and how well the owner and horse work together in completing a prescribed on-the-ground pattern. Marlon's behavior was under scrutiny from the moment we entered the arena until the class was over. After performing the pattern, everyone lines up for inspection by the judge. During inspection, you have to keep the judge in sight as he circles your horse at a fast clip. My job was to show the judge my beautiful, well-behaved horse by moving side to side in front of Marlon as the judge raced around him. When I lost sight of the judge, in desperation I'd peek under Marlon's neck to locate him. Marlon was too tall for me to see over him. This was definitely not proper etiquette for a showmanship class. Only little kids could get away with that. It happened more than once as I was learning.

The showmanship class wasn't Marlon's favorite. Practicing the required pattern, he'd do it perfectly the first time. The next time it had to be altered so he wouldn't anticipate and try to tell me what we should do. Being smart, he didn't see any reason to do a routine again if he'd done it correctly the first time. Certainly not several more times for me to learn. This totally frustrated me. After nipping at my hand, he'd throw his head up in the air. He knew he had misbehaved, and he was expecting to be disciplined.

"We've already done this." The look in his eyes told the story. He'd look to see my reaction and make sure I understood what he was saying.

Sometimes Marlon acted perfectly in the beginning, and I'd be hopeful. He picked his moment to nip at my hand. The picture we painted was not smooth and graceful. The number-one rule was for a well-behaved horse to perform in unison with you. The remainder of the class could be perfect, yet we wouldn't score at all or would score very low. Cathy and I worked and worked to correct this but were never totally successful. On a rare day, Marlon would be perfect from beginning to end, and my placing in the class would be good. It was very frustrating because from moment to moment, you never knew what he'd do next. I could forgive him the flashbacks in our riding classes, but this was on purpose, just because he could. I still had the attitude that I was the smarter one, and it was up to me to figure out how to stop his mischief. On some of his perfect days, my mistakes were too numerous to count. It wasn't my favorite class either.

Getting ready for one of my showmanship classes, I went to the tack room in our horse trailer to get dressed in my western show clothes. I met a longtime horse trainer as I was walking back to the barn to get Marlon. He had a reputation as being unscrupulous and unfriendly, and his sullen expression never changed. He did not speak to us unimportant contestants or other trainers' clients. I averted my eyes when I saw him coming toward me.

"Your hat is on backward," he muttered as he came closer.

I glanced up in time to see the slight smirk on his face. I was embarrassed at my dumb mistake and angry that this guy was the one who noticed. A lesson learned. I never made that mistake again.

The English Hunter Under Saddle class was Marlon's specialty and the reason I changed from western to English. He had the confirmation and training to win, except when the old demons came back to haunt him. It took a long time for me to build up the necessary muscles and strength to ride well. From the beginning, even when I was barely able to stay in the saddle, I loved the Hunter Under Saddle class. This class required you to walk, trot, and canter in one direction, then reverse and do all three gaits again, but not necessarily in the same order. After going in both directions, all contestants were asked to come to the center of the ring and line up. The judge walked down the line, stopped at each horse and rider, tipped his hat, and asked them to back up several steps. This was the final requirement for the class. The judge took a few minutes to add up his scores, and then the winners were announced over the loudspeaker. Waiting was agonizing. I hoped my name would be called, if only for honorable mention.

Once you entered the ring and were asked to line up on the rail, you were being judged. The horse had to be consistent and smooth. Each time the judge glanced at him, he must look the same. His top line must remain straight from his head to his hip. The horse should not raise or throw his head, and his nose should be at a right angle with the ground. The walk is a four-beat gait and very relaxed. The trot must be a long, two-beat, ground-covering stride, and the canter is a rhythmic, sweeping, three-beat gait. "He is really stroking" or "He is cutting daises" were comments often made when the horse was flowing around the arena with correct body form.

My job was to ride well and keep Marlon in a good place relative to the other horses. I didn't want him to have any distractions or interferences. The horse moved out at a fast speed, and I discovered I liked it. If I did my part well, and Marlon was listening to me, our ride was thrilling. If we did our best, I felt on top of the world. It meant Marlon and I had come a long way together. In my mind, I can replay these imperfect but thrilling rides over and over, and it gives me great pleasure.

Having a new English saddle and the necessary tack to go with it gave me further incentive to compete. I could strut around the showgrounds in my proper show clothes and almost forget I was the oldest contestant and would soon have to perform in the show ring. It was scary, but with Marlon groomed to perfection and looking so handsome outfitted in his English tack, I did feel like a contestant.

Knowing Marlon's volatility in these classes, I went around the ring talking to him in a quiet and soothing voice. "You are such a good boy today" and "Easy, easy." On and on I'd go, and it did help to keep him calm. Cathy reminded me to keep talking to Marlon.

When horses get a little nervous in a class, the first thing they want to do is have a half-hearted bowel movement. They want to stop, do their business, and then continue with the class. To go on as if nothing is happening takes lots of training. In the beginning, Marlon insisted on stopping and spreading out so a single drop couldn't splatter on his legs. He'd moan loudly and get everyone's attention. Again, only young children are forgiven for this.

During those humiliating times, I was reminded of a quote by Prince Phillip: "Anyone who is concerned about his dignity would be well advised to keep away from horses."

Years later, on a day when Marlon was calm and happy, and I was a better rider, we had a good day at a horse show. We earned two and a half AQHA points. Points are earned based on how many riders are in the class and where you placed. I had made a mistake that cost me another half a point. I kept moaning about it.

"My gosh, Nell, some people never earn a point, and you have earned two and a half points, so why are you bitching?"

Bob had put the proper spin on my situation in a hurry. It was a short time ago that I hadn't earned a single point.

CHAPTER 12
RIDING TO AVOID HITTING THE JUDGE

One weekend, Bob and I went to a quarter horse show in a small town near us. Bob wanted to take Dan and go for the day. Cathy was out of town; we were on our own, without anyone to give us advice. Marlon wouldn't tolerate Dan leaving without him, so we had to go too. I decided it was an opportunity for Marlon and me to ride around the showgrounds with other horses for experience.

We arrived at the showgrounds early in the morning, found a spot to park our truck and horse trailer, settled the horses in their stalls, and unloaded our saddles, tack boxes and all the other required equipment. We gave the horses time to explore their stalls, look around and see where they were, and take care of any personal business they might need to do. Then I saddled Marlon and rode him in the arena where the other riders were warming up for their classes. He was calm, happy, and did well as we maneuvered around the crowded arena. I was so proud of him. Enjoying my time among my fellow horsemen made me feel I belonged.

When our ride was over and we were back at the stables, Bob kept saying, "You and Marlon looked good. You need to enter the novice Hunter Under Saddle class." Having little experience, and without Cathy being present to give me guidance and support, I didn't feel comfortable entering the class. Bob didn't give up; he kept encouraging me and telling me how good Marlon had looked in our warm-up ride. Going back and forth and mentally listing the reasons why I wasn't ready, yet knowing it was a smaller show and more suitable for beginning riders, I reluctantly entered the class.

By the time the decision was made, I had to hurry to get dressed. Most shows start with the English classes, and they were first today. Our show clothes were always in the horse trailer so they'd always be present when minds were changed. I couldn't use lack of clothes as an excuse, as they had been put in the trailer for some far-off future day.

My class was called, and all the contestants entered the arena at a walk and spread out around the side of the ring. As soon as the last rider entered the arena, the gate was closed, and the trot was called for. We had a very cool judge that day; he stood one third of the way from one end of the ring and didn't move. Most judges pace around in the center of the ring with their ring steward nearby, keeping a close eye on all the horses and their riders. The ring steward passes on commands from the judge to the show announcer and keeps notes for him while the class is going on. From experience, judges know anything can happen in a novice class. The horses and riders are learning, and both are at different levels of training.

Marlon and I walked, trotted, and cantered at a safe distance behind the judge as we circled round and round the arena in the first half of the class. Standing ringside, Bob gave me a thumbs-up each time I passed by. A reversal of directions at the walk was called for. I began to worry, as this was often the time Marlon had flashbacks from his early years of mistreatment and could completely lose it. Circling behind the judge in this direction at a walk, I noticed we were inching closer to him. He cut his eyes toward me as we passed by, but he didn't turn his head or move his feet. After we'd gone several times around the large arena at a walk, the English trot was called for. Everyone moved out at a fast clip, some faster than the acceptable speed. We were moving along smartly, but as we neared the judge, I noticed we were even closer than the last time around. I asked Marlon to move back toward the side of the ring, but there was no response. Another circle around the arena, and I could see we were gradually moving in toward the center of the ring. Approaching the judge again, it looked like we were aiming straight for him. I started frantically pulling on the outside rein with all my might and nudging Marlon with my inside leg, trying to move him back to the rail. Still no response, but we had missed the judge once again. He only glanced at us as we flew by, circling behind him.

The canter was called for, and as Marlon picked up this gait, we made another big move toward the center of the ring. Right then, I started praying for divine intervention as I struggled and fought with Marlon to move him back toward the outside of the ring. He couldn't be moved. Panic set in when I pulled on the reins and kicked his side with my boot as hard as possible, and he went on as if I were just a passive passenger. This time around we barely missed the judge. He still didn't move one boot, but he saw us coming straight at him, and he turned his head to make sure we had room to pass by.

It was a tug-of-war between Marlon and me as we continued our race around the ring. Regardless of what I did, nothing changed. The next bypass looked closer than ever, and I wondered how we could manage to miss the judge one more time. I was a nervous wreck, and I could not imagine what had happened to my horse. The judge stayed planted in the same spot. Knowing he couldn't survive us circling by again without moving or running, he called for everyone to walk and line up in the center of the ring. What relief I felt! After we lined up, the judge walked down the line, tipping his hat, signaling for everyone to back, and then the class was over. I just sat there stunned and puzzled as to why I hadn't been able to get Marlon to move. This occupied all my thoughts as I walked out of the ring, and it didn't leave any room for embarrassment.

We had drawn the attention of all the onlookers at the horse show. Having problems in the ring draws everyone's attention like a magnet. I couldn't fathom what had gone wrong, and no one around me could figure it out either. It was difficult to wait to call Cathy and ask her.

When I finally talked to her and told her about my harrowing and puzzling ride, she explained, "You let Marlon drop his inside shoulder, causing him to drift in. You must keep the inside rein picked up when going in a circle to keep the horse from dropping that shoulder."

This was news to me. Soon there was lots of training for me to notice the first signs of Marlon dropping a shoulder and how to quickly correct it. In each class I entered from that day forward, keeping the inside shoulder in proper alignment was top priority. A near disaster makes an indelible impression that you never forget.

CHAPTER 13
MARLON'S BEHAVIOR, ALWAYS SURPRISING

At home, Marlon was funny, demanding, always a challenge, and oh so bossy. One reason he liked being at home was that he had someone who listened to him and responded. Sometimes the response was to his liking, and other times it wasn't. He lovingly and happily let me know when we agreed. War was often declared, as he never stopped trying to convince me to do it his way. I had to be diligent about doing everything the correct way, or in a flash he was in control. Having to be on guard every moment was such a burden. I just wanted to relax and enjoy him, but I knew what the consequences could be.

When he was happy, I loved to play tricks on him. The looks he gave me in trying to figure out what I was up to were priceless. Sometimes I walked into his stall and just stood there. Looking me over and not seeing anything in my hands, he was suspicious. When he came closer to determine what I wanted from him, if it wasn't obvious, a mild version of the look crossed his face. Wrinkles would form at his lips, and he'd hold them while he waited for me to state my intentions. His look said, "I suspect you are up to no good." In a few minutes, I had to let him know it was all in fun, or he'd become very agitated and threatening.

The horses went out to graze mornings and evenings, and Marlon was let out first. He went into the lower pastures, which consisted of two one-acre pastures with a gate between them. Before Marlon had convinced Dan that he was higher up in the horse pecking order, a pasture had separated them. Later, when Dan only challenged Marlon occasionally, they could share a fence line. We'd leave the gate open between the two lower pastures, and Marlon could go in either one. He usually went to the far end of the second pasture to graze. When it was time for him to come in, it was a long walk to go get him. I'd go to the gate, open it wide, and call. Sometimes he'd turn around so he could watch me, and he often took a few steps, but he kept grazing. One summer, I decided to teach Marlon to come when I called him. We had plenty of lush grass, and the horses were limited on how long they could graze. This didn't make it easy to convince him to come. It was a little easier in the evenings because it was time for their grain. Dan was in the pasture surrounding the barn, and he went in first. In Marlon's mind, he could be eating all the delicious grain and alfalfa hay.

One afternoon, after days of struggling to get Marlon to come, I opened the gate, called him a couple of times, and stood waiting. He looked but didn't take one step toward me. After a few minutes, I closed the gate and hid behind a tree. Danielle, who rode with us, had stopped by to visit the horses. She cut through the pastures on her way home and saw me behind the tree.

"Are you hiding from Marlon?" she asked.

I nodded my head.

"Awesome!" she responded.

After spending ten minutes or so behind the tree, I went to the gate, opened it, and again called Marlon. I waved the carrots, his coming-in treat, high in the air. Still he only took a few steps forward. I went through the same routine and again hid behind the tree. He slowly meandered forward, raising his head every few seconds, trying to figure out what had happened to me. Becoming concerned, he trotted to the gate, nickering and looking for me. This game was played over and over. Finally, he started coming quicker. He still sneaked a bite here and there along the way.

Bob laughed at me, saying, "You know Marlon. He'll outlast you."

It was time consuming. I could walk out to get him in less time. But I had to win with Marlon and prove to Bob I could do this.

One evening when Bob was bringing in the horses, I walked by a window and saw Bob hiding behind the tree. I couldn't believe my eyes after all the ribbing he had lavished on me. Marlon did come after Bob hid a few times. By summer's end, Marlon was coming when called; he was often slower than I would've liked, but he was coming. He had gotten my message.

When it was time for Marlon to go to the veterinarian, a horse show, or back to Cathy's, I'd walk into his stall with his padded leg wraps. The wraps protected his legs while traveling, especially if there were any sudden stops. A fight always took place, as he didn't want to leave home. He knew what the leg wraps represented, and it was hard to convince him he had to go. My job was to get him to stand still long enough to get the leg wraps on. When I entered his stall, he checked to see if there was anything in my hands, and if so, he wanted to see it up close and smell it. As he smelled the wraps, he'd get angrier by the minute and start complaining. I'd have to discipline him before putting them on. Hiding the wraps until ready helped to keep his anger down.

"What in the world are you two doing over there? I've never heard such a racket," Bob would say to me as Dan stood quietly for his legs to be wrapped.

Marlon was afraid to leave his home. He had moved so many times, moves that didn't turn out well. Pulling out all his ugly moves was his way to convince me to let him stay at home. I imagined I saw tears rolling down his sorrowful face. Feeling his pain, I never knew if the tears were mine or his.

After finishing, I'd let Marlon out into the pasture where the trailer and truck were waiting for our departure. He was so upset about leaving home that he wouldn't even graze the green grass. He'd hang his unhappy face over the fence and stand there until it was time to go. If we were away from home and the wraps were brought in, he'd smell them a few times, and then a smile would spread across his face. In a minute, he'd smell them again and again. After making sure he was correct, he'd stand without moving a muscle while all four legs were wrapped. Once the wraps were on, he didn't want me out of his sight. If possible, he'd follow me around. If he couldn't, he'd nicker each time I was out of sight. He was excited and ready to go home.

Despite improvement in my riding ability and in Marlon's behavior, every day was still a challenge. I had what I called "safe places" to ride in our pastures. The most level place to train a horse to make perfect circles, to stop and do 180-degree turns without stumbling, and to practice other maneuvers was bordered by a small wooded area. I could seldom make myself ride there. Dogs and critters could be hiding, run out, and spook Marlon. It had happened too many times, and I'd fallen off because of it. Once it was Grady hiding among the trees in our neighbor's pasture and giving a llama call to us when we didn't know he was there. But Diane was instructing me in my riding, and she insisted that was the place to ride.

"You do the same thing in the same place every time, and Marlon does it before you tell him," she said. "You need to set a routine before you start and follow it. Later I will draw some patterns for your warm-up and practice sessions. Don't ever stop in the same place. Vary where you turn, and make smooth transitions from one gait to the other. Don't let him lift his head or twist his body around as he turns."

On and on it went. I could tell she didn't have much patience with me for not insisting Marlon complete what he knew how to do. I admitted I was distracted by where we were riding near the woods. Again she lectured me on taking control of Marlon so that he wouldn't be looking for some reason to spook. True, but when you are scared and worried, it takes time to get to that point. I gave myself another pep talk. I was getting good at these and now called them soliloquies. Then I tried to do as she asked. She suffered through the lesson and ended it with some encouraging words.

"You've come a long way, and I'm trying to help you go further. Don't be disappointed when he doesn't cooperate. He'll have bad days like you, but when you work together, it'll make you try harder."

Days later, Bob and I were riding in our grassy pastures, and my day had been very pleasant. Remembering the lecture my daughter had given me, I asked Marlon to come to a stop, pause, then back up. Tucking his head to back properly, he noticed he was closer to the grass, and he sneaked a bite. It was so quick that I was powerless to stop it. I corrected him for taking advantage of me, but he thought it was worth the punishment. I had to keep a tight rein to prevent him from doing it again. For the rest of the ride, all he thought about was how to get another bite, and all I could do was to make sure he didn't. If he won, it'd be a battle between us every time we rode on grass.

This totally frustrated me, and I kept saying, "Will I ever get better and have the control everyone talks about?"

To save the day, I decided to do a couple of the ground patterns required for my showmanship class. This would vary today's routine as Diane had ordered me to do. Marlon didn't mind doing a couple of patterns if that was all you asked him to do. My mood lightened, and I gave it a go. His first simple routine was presentable. Feeling a bit more relaxed, I went for one more. In one swift movement, he again grabbed a mouthful of grass. He threw his head high in the air and ran backward before I could react. He knew he deserved punishment, and he expected it.

Becoming angry and disheartened, I made a stab at firmly correcting him. Then I put him in his stall and said, "Tomorrow is another day. I have done or not done all I can do today."

Every day was different, a new adventure, a new challenge, and unimaginable things could and did happen. There were few days that Marlon didn't astonish me. I was a witness to the dazzling things he did, finding it hard to believe what I was seeing with my own eyes. I never knew a horse could reason, read situations, and find so many ways to communicate with you. It was a different story when he was upset, angry, confused, or fearful. When the past haunted him, there wasn't any rational or sensible way to deal with him. I learned to enjoy the good days, when I could look into those beautiful brown eyes and see appreciation and hope from him. Those days helped me suffer through the indescribable days when he was out of control. My emotions swung from one extreme to the other. Some days I wasn't sure what to do next. A new day would dawn, and a reason to try again would pop into my head. Thinking back to the early days and knowing he needed me emotionally as well as physically, I knew I couldn't be one more person to let him down. So many people had done so.

Other days I'd say, "He has a weird way of saying he wants my help."

When becoming the owner of this flawed, intelligent, high-strung show horse, I had sensed how desperately he wanted to be understood. Marlon told me how he felt every moment we were together. Had he not been so angry, and had I not been so scared, I might not have been as observant. I kept a journal of amazing Marlon experiences and my responses to them. I jotted down notes about his actions each day and what he was telling me. He was speaking horse or sign language, but it was stated so clearly and sometimes violently that I had to know what he was telling me for my safety. I didn't want to forget a single one of his antics. In addition to being angry at the world, he was desperate for someone to understand him. He needed me to know what his life had been like and what he was thinking moment by moment. I knew how he felt. I wanted to share with someone the unbelievable things he conveyed to me. I'd tell Bob, but he was a reluctant believer at best. Cathy was the only one who understood.

"He is the most expressive horse I have ever seen, and I have been around horses all my life," she said.

I discovered he would tell me the same thing over and over until he knew that I understood. He gave me that quizzical look after telling me something to see if I was getting it. The family laughed at me when I spoke of talking to and understanding Marlon.

Bob did a lot of the barn duties: feeding, cleaning stalls, and putting the horses in and out to pasture. He had firsthand knowledge of Marlon's behavior, and he often had a different interpretation of what it meant than I did. If there was trouble with Marlon, I was called to handle it. He was my horse, and it was my responsibility to discipline, encourage, and manage him. Often the stall was the problem. Marlon had rules in his space, and you must follow them. I respected them, had patience, and chose my battles. Bob wasn't as patient. Getting too close to Marlon's feet, even touching one, or taking too long to clean the stall, Bob would be admonished. With an unpleasant look on his face, Marlon would sling his head side to side as the first warning. He was still nibbling on his hay, and he didn't want to be disturbed. Other methods would be used to get Bob's attention if he didn't heed the warning. Failing to get the desired result, Marlon would threaten to kick him. Bob would come in from the barn complaining and giving me minute details of my horse's behavior.

If our dog, Riley, or a visiting dog stopped at Marlon's stall, he'd turn his head and give them the look. Any thoughts they had of entering his stall were gone. His look sent a powerful message. The children and grandchildren were amused when I'd translate what Marlon was telling us. The only message they got was that he was a mean SOB, and he dared them to come near him.

Marlon told me when he was frustrated, and if I didn't react promptly, he'd keep repeating it. He told me minor things, like if his schedule varied or if something new had been introduced. I felt he complained if the sun was too bright. I laughed when I thought about trying sunglasses on him to see if I had guessed correctly. Sometimes I solved today's problems, and other times the war continued.

Dan was always lurking in the shadows, and some days I know he'd think, "This may be my chance to become number one!"

I was sure Dan remembered looking through a tiny crack with one eye and sending a challenge to Marlon that he wanted to be the boss. Marlon had understood and fought back in violent ways. Dan never stopped thinking of new ways to pressure Marlon for the top spot in the horse pecking order.

Marlon could carry a grudge. Once Bob gave him some bad-tasting medicine, and an hour later as Bob was leading him, Marlon bit him in the back. When Bob looked around, Marlon had backed up a couple of steps, raised his head slightly, and stared at Bob for a reaction. It was clear to Bob why he had bitten him. Marlon took a chance that he would be understood; if not, he appeared prepared for the consequences by his stance and the raising of his head.

Bob was as untrusting of Marlon as Marlon was of him. Their personalities clashed over everything. He was leading Marlon out to pasture when John, our sometimes barn helper, arrived, and Bob stopped to talk to him. Marlon waited and waited. Becoming anxious, he reached around and gave Bob a little nip, just enough to remind him he'd been standing there a long time.

"You should have relaxed your hold on the lead line and let him graze while you were talking," I said when Bob reported this to me. He had no conception of how hard I worked to give in to the small things and save the confrontations for important matters. The next morning, after the horses came in from grazing, Marlon was standing with his head hanging through the half window into the main barn, waiting for his hay. Bob reached across the divide to feel his chest and make sure he wasn't overheated. He'd seen him running, galloping, and playing with the ponies on the other side of the fence line. Marlon nipped him again. Hearing this, I went on and on about how he was invading Marlon's private space.

"Bullshit!" Bob said, interrupting me. "He wanted me to give him his hay, and when I didn't, he snapped at my arm and accidentally got me."

Admitting he had the correct grasp of the situation, I dispensed with my long psychological explanation.

Every day, my horse did unexpected things that I didn't know a horse was capable of. He was never boring, continually surprised me, and at times horrified me. Sometimes there weren't words in my vocabulary to adequately describe what he did. I kept trying to understand, letting each incident linger in my thoughts for days. I even dreamed of possible solutions. Often I had to let that incident go without an answer as the next mind-boggling one was occurring.

Discussing this with Diane, I admitted I loved a good challenge, just not such an impossible one.

"I know you do. After all, Dad's a challenge," she quipped.

I was stunned for a moment, then had to admit she was right. He certainly had his high-maintenance times.

Besides the frustration, misery, horror, fright, and too many other feelings to name, having horses provided real-life entertainment. One day we discovered Eli, one of two ponies that shared a fence line with Marlon, was in Marlon's pasture, just walking around exploring. Suddenly, he was seen by his disbelieving neighbor, who was appalled that someone had dared to enter his territory. Marlon lowered his head in a threatening way and charged toward him. Eli ran for his life. He didn't have time to fly through, over, or under the barbed-wire fence. He zigged and zagged in every direction. In a brilliant move, he edged in close to the outside of the round pen, making it easier to maneuver the turns. Round and round they went. We could see Marlon getting angrier each time he raced around the pen, and the pony was still there.

Bob and I had seen the little intruder in the pasture first. We couldn't think of any helpful thing to do. Didn't have a clue how he had arrived. We just watched and let them resolve it. Getting nervous, we couldn't see how this could end without one or both getting hurt. Then, quick as lightning, Eli made a dash for the wire fence. He hit it full force and landed on the other side with only a slight hesitation as he dragged his hind legs through the fence.

Shocked, Marlon stopped in his tracks. He stared in disbelief. Watching, we felt the same way. Now that it was over and had ended well, we could relax and laugh. Laugh we did, for hours. A word about Eli from either of us would start the laughter all over again. If having a horse is your hobby, it's wonderful to have someone to share it with. No one else would believe all that goes on between you and your horse except someone who witnesses the unbelievable with you. Besides needing someone to laugh and commiserate with, you certainly need someone to pick you up when the falls occur!

A few months later, going to the barn, I saw a horse in the pasture with Marlon, slowly walking toward him.

Remembering the time with Eli, I was thinking, "Oh, no! This can't be happening again. He doesn't allow horses in his space. He barely allows the people who care for him."

In spite of my fear, I smiled about Eli's ordeal, but this was a full-size horse. Looking, I recognized Satin, recently acquired by the neighbors. Marlon quit grazing, raised his head, and stood still as he watched her approach. They touched noses, turning their heads to the left and right as they did so. I discovered I'd been holding my breath. They stood together communicating for a long time, then lowered their heads to eat side by side. Seeing some of his good points, it reaffirmed why I forgave him again and again. Satin was twice the size of the pony and was in Marlon's pasture, but she was female, and poor little Eli was male.

Dan was in his pasture sharing a fence line with Marlon when he saw Satin and raced to the fence. Marlon and his lady friend flew to join him, and the squealing, biting, rearing up on hind legs, and fighting began. Jealousy had kicked in. I rushed to catch Dan and put him in the barn before deciding what to do about the visitor in our pasture.

"Your horses live better than a lot of people," our vet often said.

Dan and Marlon did live well. Many show horses lead sheltered and pampered lives. They get used to it and then expect it. Fans were added to the stalls to keep them cool in hot weather. We put our horses out to pasture early in the morning and late in the afternoon so the sun would not bleach their hair. Before going outside, they were sprayed with fly spray, and fly bonnets were put on to protect their eyes and ears from gnats, bugs, and flies. They understood what these were for and lowered their heads so we could put them on. Seeing the horses with the fly bonnets on, some people thought we had blindfolded them. But the bonnets were made from a very fine netting, and the horses could see perfectly.

As much as humanly possible, we kept Marlon's schedule the same; otherwise the price we paid was too high. He would get so worked up that he'd race around the stall at full speed and kick the walls. The running in the stall made the biggest mess, grinding the manure into the bedding and kicking sawdust under the heavy rubber floor mats. We'd have to wrestle the mats back and level the sawdust underneath before putting the mats back in place and covering them with a thick layer of sawdust. If the flooring was uneven, it was easy for the horses to pull a shoe. If this happened, they couldn't go outside, and there wasn't any riding until the farrier came and replaced the shoe. Marlon's walls were now covered with heavy rubber in the most amusing pattern, but I still worried he'd become lame in his kicking leg. He continued to use that right back leg to express himself, and often the ankle was swollen.

Besides the blankets and hoods, heat lamps were hung in their stalls to keep them comfortable in cold weather. Spotlights were turned on for sixteen hours a day to keep the horses from growing winter hair. Putting the blankets on and off caused static electricity. When this happened, Marlon behaved as if I had shot him with a gun. I solved the problem by rinsing the blankets in plenty of fabric softener after washing them.

We usually brought the horses home in late spring and kept them until bad weather arrived. The remainder of the year, we boarded them at Cathy's barn. Marlon couldn't wait to get home from wherever he was. Dan never seemed to worry; his personality was laid back and calm. He had never faced starvation or abuse, so he wasn't nervous about where he was going.

With the horses' limited time out to graze, there was seldom a manure pile in our pastures. Friends visiting would admire the well-kept pastures with the pretty horses out grazing, then notice the lack of manure piles.

"How can your pastures be so clean with horses out grazing?"

We explained that Marlon and Dan knew their schedule, and they went to the extreme to make sure they urinated or had a bowel movement before going out. When they saw us coming, it reminded them of what they needed to do, if they hadn't already taken care of it. Smart horses want to go in the deep sawdust instead of on the hard ground. We patiently waited while they prepared to go out.

When the horses were at Cathy's, they were happy to see us, unless Marlon was angry over how long it had been since the last visit. Recognizing the sound of our truck, the horses called us before they could see us. As we entered the barn, they would be standing with heads held high and ears forward, anticipating our arrival. Loud neighing bombarded us. We'd rush over and greet them royally. These were happy times. If Marlon was unhappy with how long it had been since my last visit, he'd be lined up along one side of his stall with his head in the back corner. He wouldn't turn his head and look at me as I walked back to see him. Talking to him in my happiest voice, I'd be ignored. I got his message that today there would be trouble between us. The back corner of his stall was his place to shut out the world, to nap, to rest, and, at times like this, to pout. When I'd get near him, he'd turn his head, look at me, and give me his most neglected, pitiful look. He was such a master at this. Laughing, I'd hug him and keep telling him how much he was missed. It varied how long it took to improve his mood and get him to take his place at the front of the stall. If Bob walked by, Marlon would tell him to keep on going. He was enjoying his special time, and he didn't want Bob to interfere.

The times he was angry at barn problems, he wanted me to know and sympathize. His emotions ran the gamut from one end of the spectrum to the other. He'd be standing in his favorite place at the front of his stall with his head hanging over the half door into the barn hallway. Seeing me, his ears would shoot straight up, giving the impression they reached to infinity. He alternated his look between letting me know he was happy to see me and wanting to tell me his latest problem. The wrinkles would start to crawl up his face, then quickly disappear. His angriest look could appear in a second. If it was minor, I could have him in a good mood in minutes. Major problems could keep him angry to some degree for my entire visit. A new horse across the hall, or a horse in the next stall doing things he didn't like or understand, could be considered major.

When I talked to Cathy about Marlon's many moods, she told me about the day the farrier had come to put new shoes on Marlon. He had a sinus infection, and after he'd been bending over a short time to work on Marlon's feet, he'd stand up, cough, clear his throat, and loudly blow his nose. Marlon turned his head and looked at him each time he stood up, as if to ask, "What is wrong with you?" It took a long time for the shoeing to be completed. Marlon didn't seriously complain, but he never stopped questioning what was going on.

The next day, the farrier returned to shoe other horses. Every time Marlon saw him, he'd shake his head, put his ears back, and convey to him, "Don't come near me today."

At the end of the day, as the farrier was leaving, he couldn't resist walking by Marlon's stall to say goodbye. Marlon let him know he still wasn't happy with him.

Cathy had individual turnout pastures. Each day, the horses took turns going out for several hours to run, play, and graze. A retired pony stayed in one pasture, and when it was Marlon's turn to go out, he'd stop walking in front of the gate and beg to join Danny. They had great fun playing together. If Cathy didn't let Marlon join Danny, he stayed angry with her all day and reminded her each time he saw her.

Show horses are normally in a corral or pasture alone. At different times of the year, they wear blankets, hoods, or fly bonnets, and another horse can rip them off, destroy them, or injure the horse. They can fight, bite, and kick each other, and this often leaves scars on their beautiful coats. This is not desirable for show horses, especially during show season.

Sometimes Marlon would be having his little stand-up daytime nap when I arrived. He'd give me mixed signals, saying, "I'm happy to see you, but please let me finish my nap." He ate his grain early in the morning, then nibbled on his hay until it was time to go out to graze. By the time he came in from grazing, he was feeling full, lazy, and sleepy. Dan lay down and slept. My tense horse only napped standing in his corner. He needed to be ready to move quickly if a need arose, and he was sure one would. I had to talk tough to convince him I could only stay so long, and he would have to adjust.

CHAPTER 14
CHURCH, THE COWBOY WAY

Horse shows are normally scheduled for the weekends and run for three days. Most participants work during the week and can only show on weekends. This is a problem for churchgoing cowboys and cowgirls, so a group of them got together and started the cowboy churches. They were just beginning when I first started going to horse shows.

Singing could be heard throughout the showgrounds early on Sunday mornings, reminding the cowboys and cowgirls to come on down for the church service. It was held in the bleachers along one side of the riding arena.

It was difficult for me to attend the church service. My classes were early in the morning, and I had to be on the showgrounds by five o'clock or earlier, if someone didn't feed for me. My uptight horse was antsy in a new place with so much activity. After eating, he needed to be exercised. I hated this because it was so time consuming. When I'd think he was finished, he'd leap into the air and start running again. It was a balancing act to determine when Marlon had calmed his nerves yet had the right amount of energy and stamina left for our classes.

After the exercising marathon, my dear buddy did not want to cooperate with me as I groomed and saddled him. I was frustrated and in a hurry. Finally, I'd tie him up with a loose rein so he could nibble on hay but couldn't lie down. I hurried to get to the church service and managed to get there most Sundays.

A couple of years earlier, at the first service I ever attended, I rushed in after the singing had started. Looking for a place to sit, I saw an empty space at the end of a bleacher next to a cowgirl.

"Is this seat saved?" I asked.

"Yep, I sure am!" she said, looking up. I realized she thought I was asking whether she was a Christian and had been Saved.

"Me too!" I replied as I sat down by her.

I was thinking about that as I went to this cowboy church service. Arriving, I sat down and looked around. The first startling thing I saw was a lady climbing the bleachers with a large German shepherd. The dog sat down beside his owner on high alert, staring at everyone as if inspecting them, barely touching his seat and staying ready to jump into action.

"Is this a security dog?" I wondered. "If so, why is it needed?" As I was contemplating this, a fight occurred between two dogs as their owners approached the area. No harm was done, but there was a lot of noise. Glancing around, I saw there were almost as many dogs in attendance as there were people.

Cowboys and cowgirls were riding in the arena as quietly as possible on the far side of the ring. The footing was good in the arena, so a few people were lunging their horses.

This isn't a quiet activity as the horses gallop, kick up their heels, and squeal.

I was used to these activities and noises, but sitting there, it seemed surreal. I felt like laughing out loud at the sheer lunacy of it all. Before this feeling could overcome me, the singing ended, and a gentleman gave a short message. Some riders on horseback came and stood nearby so they could hear. Then a cowboy and his wife sang a duet. The audiences said amen. Next the cowboy read a passage of scripture and gave a sincere testimony, a benediction was prayed, and the service was over. I felt it was short and meaningful.

I hurried back, hoping that Marlon and I would have enough riding time. We needed to get in sync, and I wanted to impress upon him that today I would be in charge. Also, at this older age, my muscles required longer to warm up. I never knew how long it would take for me to feel that we were as ready as we could get.

Later that afternoon, after I had finished my English classes, Bob and I were ringside watching some of the western classes. As one rider approached, I could hear him cussing his horse. In a low voice, he was calling it an SOB and other names. This was most unusual. The next time he came by, I looked closer and could not believe that it was the singing cowboy who had sung and spoken at the earlier church service. I was shocked! Cowboys will cuss their horses, but they become perfect gentlemen when they enter the show ring. They tip their hats to the judge and only speak in a polite, soft voice if spoken to.

"Well, I have seen and heard it all today," I told Bob.

"You know they are barely adequate at best," Bob replied, summing up the cowboy services.

This happened in the early days of starting cowboy churches at horse shows. Over the course of a few years, the services became better and better and grew larger as word spread.

CHAPTER 15
MARLON BOLTS, AND I AM THROWN

I fell off Marlon seven times but was only thrown once. Each time Bob and I rode, we worked hard, trying to remember and practice what Cathy had taught us. Some days we were happy, thinking we had made great progress. Quick as a flash, something would happen, and the good feeling would disappear. For me, it was such an up-and-down emotional challenge. The tiniest bit of progress could put me on a high, but if I didn't do my best, or Marlon pulled one of his antics, I quickly dropped back down. Such extremes were tough to take. On the days I fell off, it was difficult to keep my confidence up to try again and again. Deep down I knew each fall could be my last but was never scared enough to let that dominate my thinking. After recovering from hitting the hard ground, my thoughts would begin to soar once again, and I'd think, "This too shall pass, and I will survive."

On a fall day, we went to Peachtree City, Georgia, to see our granddaughters play soccer. It was a four-hour drive, and the stable was on the way. The soccer games were scheduled for late afternoon, so we decided to leave early, stop by the stables, and ride the horses.

Driving to the stables, we noticed the wind was exceptionally high. Bob said, "What's new? The wind always blows on Sand Mountain."

We arrived at the stables, greeted our buddies, saddled up, and joined Cathy and several of her clients in the outdoor riding ring. Walking out to the ring, we noticed the wind was rapidly increasing. I was still reluctant to ride on windy days, but if I didn't get out there, I'd seldom get to ride. There wasn't an indoor arena; Cathy had recently moved to her own place.

In spite of the wind, Marlon and I were doing well and enjoying having friends to ride with. We had been riding for about an hour, and it was almost time for us to leave, when the unthinkable happened. Cantering close to a small bridge surrounded by a blue tarp to imitate water, we heard a rattling noise. Suddenly the wind raised one end of the tarp high into the air directly in front of Marlon. For months, the tarp and bridge had been there for the trail riders to practice walking over, and nothing had moved. Marlon swerved and bolted, tossing me through the air like a rag doll. I landed on the ground before I was aware that I was falling. Lying there stunned and hurting, I heard footsteps approaching and voices talking to me.

All the other falls had felt like they were happening in slow motion, and I was aware of each second. This time was different. There wasn't any time to be afraid, to react, or try to break my fall. Wham, bam, I was on the ground! I had landed on my right hip. A second after my initial hit, my head hit the ground. Realizing I'd suffered light whiplash, I had an epiphany at that moment to never ride again without a hard hat securely on my head. I should have had one, but western riders didn't wear helmets. After changing to English, it had never entered my mind or Cathy's to start wearing a riding hat.

Then came a moment of panic. I had to check and see if anything was broken, and if so, at my age, could I ride again? I was dazed and in pain. All the riders were trying to help me and evaluate how badly I was hurt. Being thrown from Marlon's height is difficult to survive without serious injury. My hip hurt, and I lay there waiting for the pain to subside. Finally, I got up and hobbled around, trying to assess the damage. My hip was bruised but not broken. I had a very sore finger. The small buckle holding the reins together had ripped through my hand as I was thrown through the air. After calming down and realizing I had survived another fall without a serious injury, I noticed Marlon standing a few feet from me, shaking uncontrollably. Rushing over to comfort him, I saw that his whole body was quivering. The look in those expressive dark eyes told me how terrified he was. We slowly walked around so I could reassure him that we were both OK, and he wasn't at fault for what had happened. Knowing he felt he would be blamed, I needed to keep telling him all was well. Everyone was still watching and waiting for my next move.

"Guess it's time to get back on my horse and finish my ride," I said, letting them know I might be a bit frayed and frazzled but was able to carry on.

Even though the ride wasn't long, Marlon and I needed time to check our physical and mental condition. Riding kept Marlon's mind occupied while he calmed down, and it gave me time to sort things out.

After finishing at the barn, we left for Peachtree City. When we arrived and got out of the car, I was stiff and sore from sitting so long after the fall. Moving around helped, and after a late lunch, we all went to the soccer games. Friends of Rob and Karen were sitting with us. Later, standing up to walk and stretch my legs, they saw me struggling to get up and asked what was wrong. Karen explained what had happened to me earlier that morning.

"Hey! I guess no shuffleboard for you two?" one of the friends asked.

Rob thought Marlon was a mean, unpredictable, unsafe horse for me, and this was a warning. He had been concerned from the beginning, and this incident reinforced it.

"Hey, Mom, eat all the junk food you want, because it isn't going to matter," he told me later that evening.

My family was encouraged to eat healthy. It was a family joke that they were caught when they ate high-fat junk food. His meaning was clear: I wouldn't live long enough for unhealthy eating to matter, because Marlon would take care of it. He was teasing me, yet he felt there was truth in his statement.

CHAPTER 16
MY GRANDDAUGHTERS

My four granddaughters are called my angels. They arrived two by two, one from Rob and one from Diane; a second pair came along three years later. Close in age, they loved to play together and seldom needed me to referee.

"You have it made. You have four granddaughters that get along, can entertain themselves, and love to come to the farm," Diane said to me one day when we were discussing them.

Julianne belonged to Rob and Karen and was the oldest and the leader. She was on the quiet side and very thoughtful. When she spoke, the other girls listened. Jessica, four months younger, had a sweet, permanent smile on her face and loved to play pretend. She was Mike and Diane's firstborn. Three years after Julianne and Jessica joined our family, along came Jeannene (Julianne's sister) and Rebecca (Jessica's sister). One day, the four girls were visiting us and playing in one of the bedrooms. I hadn't heard a sound from them, so I went to check. Jessica was curled up in the back of the closet, and the other girls were sitting in the doorway staring at her.

"What are you doing?"

"Nothing," they all said at once.

"What's going on, and why is Jessica curled up in the closet?" I again inquired.

"She is Perdita, and she's having puppies," one of them finally replied.

I wished her well and left with a smile on my face. We had recently been to see the movie *101 Dalmatians*.

Jeannene was tall for her age, athletic and the tough one. She could fall, have bloody knees or elbows, and leap up and keep going. Her sister and cousins didn't like to see blood. Rebecca was seven months younger than Jeannene. Growing up, she was the shortest; the other girls towered over her, but she was spunky and a deep thinker. She kept us on our toes, asking questions without answers. At three years old, she found gymnastics and didn't have as much time as the other girls to perfect being a cowgirl.

Julianne and Jeannene came to visit as often as they could. They lived in Peachtree City, Georgia. Jessica and Rebecca lived across the pastures from us. It appeared to the out-of-town girls that we were constantly having fun on the farm, and they were missing out. When they came to visit, we made up for lost time. The dogs had to be trained, and the "pretend" horses had to be ridden until the girls were old enough to ride the real ones. When they could sit in the saddle on a horse, they were thrilled. Mama had a new job: I had to walk or jog alongside Dan in heavy sand to make sure the girls didn't fall off. Diane lived here and took her turn. I told Rob he owed his mom for all those miles I jogged in deep sand.

A few years later, when the older girls were eight and the younger ones were five, Marlon came into our lives. Going to the barn with me, the girls observed all the scary and unusual things Marlon did and how I reacted. I tried to explain his behavior, but so much of the time, I didn't have an inkling of what he was up to. As time marched on, I began to understand more of what my horse was telling me. I advised the girls to stay a safe distance away and observe, then translated what Marlon was saying. Watching, they never made a sound; afterward, they bombarded me with questions.

When Marlon was angry, he'd kick the wall as hard as he could and startle the girls. After jumping, they could laugh with me. Later, when their parents asked them what they had been doing, they gave long and complicated versions of what Mama and Marlon had done and said. Lots of discussion ensued. The more talking and laughing that occurred, the greater my and Marlon's reputation grew.

One or more of the girls were often around when Bob and I were saddling horses or working at the barn, or they rode with us in the pickup truck. One day as Jessica, Papa, and I were driving down the highway, Bob and I ended up in a lively argument. We weren't angry, but each of us knew we were correct. We raised our voices in the cab of the truck to make sure the other one got our point. Looking over at Bob while rebuking his latest comment, I noticed Jessica looking very concerned.

In my grandmother voice, I asked, "Do you know why Papa and I argue?"

She shook her head, indicating she didn't have a clue.

"We are circulating the blood to our brains!" We all laughed.

From that day forward, if we barely raised our voices, the granddaughter present would look at us, laugh, and say, "Are you circulating the blood to your brains?" This let them know everything was good with Papa and Mama.

Years later, when the girls were older and Marlon was calmer, they started learning to ride him. Suddenly, they understood some of what he was telling them. They took great pride in learning to ride, having patience, and being able to communicate with the horses, especially Marlon. He was the ultimate challenge. If he was having a bad day, there wasn't any riding. We needed him to cooperate as much as possible. Special memories were made, with stories told of their growing-up years with Marlon and Dan. Marlon provided the largest number of remembrances with his background and temperament.

As the girls learned to ride, they each had the opportunity to enter a walk-trot class on Dan when they felt they were ready. It was a challenge for each girl to get good enough and have the confidence to enter the class.

After Julianne, the oldest, had her turn, she composed this poem about how she felt.

> "Today Is the Day"
> Today is the big day for me.
> The day I show for the first time.
> My stomach does flips,
> But the horse is as calm as can be.
>
> My mouth is dry from all the dust,
> Sweat is pouring down from my head.
> It is so hot!
> But I have to show.

I finish dressing
And go check on my horse.
Then I climb aboard.
And here we go!

Around the arena
At a walk,
And now a trot.
"Come on, boy, don't stop."

Finally, I'm done.
And I come out with my goal completed,
A blue ribbon in my hand.

In another one of her poems, she described her grandmother with this verse: "She is not the best rider yet, / But she tries very hard. / She is a good teacher, / And teaches me things I never knew."

When Julianne and Jessica entered college, and Jeannene and Rebecca were in high school, it was harder to schedule visits to the farm. They still wanted to hear the latest news about me and Marlon. Marlon had recently gone through a very traumatic experience. I knew he had to send a letter telling them about it.

> Dear Angels,
>
> You know that horse shows cause me a lot of stress under the best of circumstances. We

went to one a couple of weeks ago. I hadn't been in a long time, but I wasn't too unhappy. Arriving at the showgrounds, along with lots of trucks pulling big trailers loaded with horses, I began to feel anxious. After unloading from the trailer and walking down the barn hallway to my stall, I couldn't help noticing the other horses were watching me. Trying to calm me, My Lady put hay in my stall for me to munch on. The horses continued to watch me while I nibbled. I do not like new horses challenging me when I can't respond. Throwing my head from side to side, showing them my ugly face with my ears flat back, was my way of objecting. When that didn't work, I tried kicking the wall with both back feet as hard as possible. This upset My Lady. She rushed in and wrapped my back legs with the big fluffy wraps so I wouldn't hurt myself. Don't know that it helped! Sure didn't slow me down! It did make her a little happier.

Oakey, a friend, arrived and was put in a stall nearby. Knowing who is the boss, he didn't challenge me. Dan, my good buddy, was across the hall. He was behaving himself because he was very happy to have me nearby. When we're away from home, we are each other's security blankets. It was a busy and unsettling time, with people and horses arriving throughout the night.

Early the next morning at feeding time, the horses near me went right back to their former shenanigans. They stared at me, jumped up and down, kicked the walls, and raced around their stalls. This really made me angry. It was an all-out kicking war. Cathy saw everything but didn't tell My Lady. It was show day, and she didn't want to make her nervous.

After being lunged, which gave me plenty of exercise, Cathy rode first. I settled down, forgetting my anger and nervousness. She bragged on me, saying, "An excellent workout, sweet boy."

My Lady groomed me, beautified me, and saddled me up, and we went for our warm-up ride. Both of us relaxed, had a good ride, and felt ready for the day ahead.

Class time arrived. We entered the indoor arena with a large group of horses and riders. The trot was called for immediately. It's a fast-moving gait that spreads the horses out and helps us prepare for what lies ahead. I did excellent in the first direction for all three gaits, walk, trot, and canter.

The judge called for a reversal in direction at the walk. We walked and walked, circling the big arena. I was so good! My Lady kept telling me what a good boy I was. We kept walking, and I began thinking, "Enough already! Get on with it, Judge!"

Another trip around the arena at a walk and I knew I had to move this class along. So my downfall began! Giving a shrill cry, I leaped forward and began to canter in place. My Lady was trying to hold me back while firmly saying, "No! No! Walk. Walk."

I just lost it completely. So sorry to say but the long last half of the class found me doing everything I wasn't supposed to do. Running in place, throwing my body around like a maniac, kicking my hind legs up in the air, and trying to gallop. My Lady kept trying to get me back on track, but it was hopeless. I wanted to race around that ring as fast as I could go and finish this class.

When the judge finally called for the canter, I squealed and thought, "Oh boy, here we go!" Tried to take off. Would you believe My Lady wouldn't let me go? She wrestled me back to the rail and made me wait until she was ready for me to canter. When she finally gave me permission to go, I was so thrilled I leaped out on the wrong lead. In my excitement, I temporarily forgot you always lead with the legs facing the center of the ring. With the speed and struggle going on between us, My Lady didn't notice. Cathy was standing ringside, and as we passed by at full speed, she called, "Wrong lead."

Immediately, My Lady asked me to change leads. This was an exciting call for me, not having to slow down to a walk to make the correction. I could change to the other lead as I flew through the air. All my pent-up nervousness just exploded. It was impossible to slow me down now. Everyone was watching to see what would happen next.

The good judge noticed how out of control this horse was and signaled for the announcer to say, "Walk and line up in the center of the ring." Thank goodness this class was finally over.

As we were walking past the judge, My Lady told him, "Boy, this horse was in top form today." The judge defended me, not believing a horse could act as I did without a legitimate reason. "Something must have scared him down at the other end of the ring," he said.

My Lady knew better, and she was ashamed for both of us. What can I say? When I'm wrong, I am totally wrong! She always forgives me and says she is supposed to be the smarter one; therefore, she needs to come up with a solution. Knowing how disappointed she is in me, I will try to make it up to her.

Hope your day has been better than mine.

Love, Marlon

A few weeks later, I received a message from Julianne. "Please tell Marlon my mailbox is empty."

Marlon was happy to have someone besides me to tell his troubles to.

CHAPTER 17
FIFTIETH WEDDING ANNIVERSARY

When you are busy, the years seem to fly by. In the fall of 2003, we realized our fiftieth anniversary was fast approaching. I chuckle anytime I think or say we've been busy. It reminds me of the time Bob and I went to the vaccination clinic to get shots for a trip to Thailand. After the nurse gave them to us and we started to walk out, she called to us in her heavy German accent. "Oh no, you cannot go! You must wait fifteen minutes."

With a slight frown, Bob immediately held up his arm so that he could see his watch. His look said it all, even though he didn't say a word.

"You retired people are the busiest people in the world," she said.

I often wonder if we are the busiest or just the most impatient group.

We couldn't decide on how to celebrate our upcoming anniversary. Every time we brought up the subject, we just couldn't wrap our minds around the fact that it had been almost fifty years. We'd end up saying, "Do you remember when this or that happened?"

We could be watching television or snuggling in bed preparing to go to sleep when one of us would mention something from the past. The television show was forgotten, all thoughts of sleep abandoned, and we would talk and laugh for hours. We marveled that we were still together and more in love than ever after the turmoil we'd endured from our families in the beginning. We were so young, but determination and love had carried us through. Becoming parents of a boy and then a girl was so exciting and thrilling, even without any experience. Poor Rob; we practiced on him. It seemed that only a few years later, we were grandparents, and again our world changed forever.

Our discussions covered the wonderful standout events as well as some of our knock-down and drag-out word fights. We could talk about the fights because we had long ago forgiven each other. The angriest I ever became at Bob was when he wanted to get a motorcycle. He was seventy years old and had a horse requiring all the physical stamina he could manage. He hadn't done too well when we tried out bicycles. After many angry words thrown at each other, he won the battle. He took a safety course and then bought a multiple-gear "hawg." I felt a certain amount of satisfaction when he couldn't keep up with which gear it was in, and I thought he'd sell it. To my dismay, he traded it for a huge, road-suitable motor scooter. I was scared for his life, but he made it through another old-age crisis. Bob just smiled as he listened to me, and he agreed he'd never seen me angrier.

Then he said, "Since you brought up being angry, I found it incredible that as soon as we bought a house and planned on retiring in this town, you wanted to buy cemetery plots. Of all the things you could have bought for a new home, cemetery plots were at the top of your list. We were in our forties, and I suggested we had plenty of time to do that, but your 'always be prepared' motto had kicked in, and after many loud, angry discussions, I knew I had to give in. I am happy to say we haven't needed them yet."

We spent a lot of time going over the twenty-nine years we had moved around the country and beyond with the military. The moves were both new adventures and sad times for us, as we had to leave our friends and all that was familiar. Every time we laughed or cried over an event, we'd end up saying, "That was a great year!" On and on it went, one good year after another. Soon we noticed we said that about all the years. Even when bad things happened, we had each other, and we overcame and survived. Time had mellowed the hard times and heightened the remembrances of the good times. We appreciated hearing each other express such feelings as, "You are as pretty/handsome as you were the day we were married, and I love you more than ever!"

We finally decided to celebrate our anniversary with a dinner party for relatives and close friends. We chose a lovely garden room at an Italian restaurant for the occasion. The granddaughters selected music to play throughout the evening. It was an eclectic group that gathered. Young and old, they represented people from different walks of life that we had encountered in our many adventures over the years. As everyone mixed and mingled, the conversation was loud and lively. When dinner was announced, we sat down to a great Italian dinner, a few champagne toasts, and further visiting with each other.

After dinner, Rob stood up and welcomed everyone on our behalf and announced he had composed a "top-ten list" of why he thought his parents' marriage had lasted for fifty years. Bob and I looked at each other and thought, "Oh no!"

TOP TEN LIST

10. Make sure she's not marrying you for your money by not having any.

9. Don't laugh at your new bride when she doesn't even know how to make a peanut butter and jelly sandwich.

8. Have two nearly perfect children.

7. Don't question your husband's masculinity when he brings home a hot-pink scooter to ride around at the horse shows.

6. Don't laugh when he later overcompensates by buying a "hawg" and leather jacket.

5. Make sure you don't marry another crossword puzzle addict.

4. When your wife leaves the room, make sure there is always a dog or small child nearby to blame things on.

3. Don't argue over whose horse is better or smarter.

2. Drive separate cars to all social events.

1. Don't waste money on no stinkin' blood test. Elope to Mississippi!

Bob and I were further surprised when Diane got up and said, "I can't let my brother outdo us. Mike, Karen, the four granddaughters, and I have composed a song telling their story."

Mike played the guitar, and the girls sang.

> "Strawberry-Picking Queen of Sand Mountain"
> Gonna tell you a story that you won't believe
> How Bob fell in love studying chemistry
> With a girl from Sand Mountain, who wanted to be free.
>
> CHORUS:
> He fell in love with a strawberry-picking queen (run, Bob, run; run, Bob, run)
> Sweetest little woman anybody ever seen
> Special-lee on Sand Mountain.
>
> He said, "Let's marry,"
> She said, "Now
> We're gonna get married some way somehow.
> Let's find a place where we can exchange our vows.
> It's now or never, don't run away;
> Let's go to Mississippi and elope today."

Tonight's the night; Bob will finally get his way.

CHORUS

Bob had often told Rob, Diane, and the granddaughters that he'd married the strawberry-picking queen of Sand Mountain. They had lots of fun composing many verses to this song and singing it to the tune of "Roller Derby Queen," by Jim Croce.

Listening to them sing transported me back to those strawberry-picking days. I could still see the disbelief on the faces of the pickers as the number of quarts picked by each person was totaled at the end of the day. I couldn't believe it turned into a contest that I happened to win.

I was fourteen or fifteen years old when the Crye Brothers advertised they had planted acres of strawberries on their farm. Word spread quickly throughout our farming community that workers were needed to pick the strawberries. Notices were posted at grocery stores, barbershops, churches, and any place where people gathered, saying, "A flatbed truck will pick up workers at designated locations on strawberry-picking days. Times will be posted for when the workers are to be at each place." Most able-bodied people were interested.

Growing up on our family farm, I did every kind of work there was. When we hauled hay, I was positioned on top of the wagon to level each forkful as it was tossed up so we could haul a larger load. Cotton was picked in the fall; for incentive, my brother, Bobby, and I were paid a penny a pound. We worked hard and counted the pennies we made and dreamed of all we could buy. But there was a catch. We had to use the money to buy our needed fall and winter school clothes. We seldom made enough money to cover the necessary socks, shoes, and underwear needed for the year. At times we were discouraged, but it helped to dream of possibilities while looking at the Sears and Roebuck catalog.

Corn was grown for people and animal food and for emergency cash during the year. We had two mules for farming; one cow for milk and butter; a bunch of chickens, mainly for eggs; and a couple of pigs for meat and lard. I fed the animals as needed, and when my mother became allergic to handling milk products, I was the one chosen to milk the cow. My dad had taken a job working as the bookkeeper at the local cotton gin. He had to be at work by four in the morning, as the farmers came early with their loaded wagons.

Before I could milk the cow, I had to practice and build up my hand muscles. I didn't have my heart in this job, and practicing went on and on. I assumed you caught hold of the cow's teat, squeezed, and the milk came out. I had to learn to grasp the teat with my hand and apply pressure firmly from the top down. If you didn't do it correctly, the cow was unhappy, and she let you know by firmly swishing her tail. The ends of the tail often slapped you across the face and could hit you in the eye.

I despised this job for more reasons than one. Our barn was across the highway and a side road that intersected in front of our house at a thirty-five-degree angle. To milk the cow twice a day, I had to cross both roads early in the morning and return late in the afternoon dressed in my old clothes and carrying my milk bucket. Traffic volume was low, but it seemed to pass by at these times. My challenge became to hide and watch until there wasn't any traffic coming, then make a run for it. It was difficult to run carrying a bucket full of milk.

Of my many jobs, I had never picked strawberries. But I wanted and needed to make some extra money, and there were few opportunities. When hearing about the strawberry-picking job, I was excited and knew it would be interesting to work with other people and to do something different. This wouldn't be another lonely farm job; I would have company while I worked. I could only daydream so much.

Our house was between designated stops to pick up workers on strawberry-picking days. The driver made an extra stop for me. Teenagers as well as adults were interested in this opportunity. We sat around the edges of a huge flatbed truck with our legs and feet hanging over the sides, talking and laughing. A man was posted to the back of the truck to keep order and to keep us safe. When we arrived at the field where we'd work, people were already there waiting for instructions. We were told to pick up a wooden crate with a handle that held twelve quart-size boxes. After we filled the twelve boxes with strawberries, we were to turn them in to one of the field stations scattered all around.

At first, people were milling around, talking and figuring out their plan for the day. When we were told to spread out and start, I went to work. I had no idea how long it would take to finish the field, and I wanted to pick as many quarts as possible.

Every time anyone passed by, they'd ask, "How many quarts have you picked?"

As the day went on, word spread that I'd picked the most strawberries. I became aware of this when we finished and were being paid. I was so far ahead that some people wanted to know how I did it. This was baffling to me. All I did was work as fast as possible to earn money.

A few days later, we were back in the field picking strawberries. I was greeted with, "I can't believe you picked so many quarts of strawberries last time. How did you do it?"

Several of the younger people announced, "I am going to pick the most strawberries today."

I have always liked a good challenge, but this one was so unexpected and not one of my choosing. The race was on whether I liked it or not, but I would not give up without trying. There was a constant stream of people coming by to see how many quarts I had filled. I didn't ask about the other pickers, but they passed on information about them. I just kept on picking strawberries, regardless of what was happening. At the end of the day, I had again picked the most strawberries. From this point on, the comments by those trying to outpick me became a bit nasty. A few people accused me of having someone helping me on the sly. People chose sides. Some encouraged me; others were determined to pick more strawberries than I.

The community was aware of the competition, and it dominated the conversations. Finally, the strawberries were gone, and the game was over. I had picked the most quarts every day, and this landed me the title of strawberry-picking queen. I was so happy it was over. Crawling around on the ground or squatting for hours at a time on dry or wet ground was hard work. Also, I could never relax or talk to friends until the day's picking was finished. Even then, it was the same old subject. Early on, I figured out why I was winning. I picked with both hands continually, freeing one hand only long enough to move my container. Most people held the quart container in one hand while they filled it with the other.

"This is the way hardworking people have some fun," my dad said, putting the contest into perspective for me.

The song went on and on, verse after verse, covering so many events of the past fifty years. I was amazed that my children had heard about these things, as I wasn't one to talk about the past. The strawberry-picking queen hadn't crossed my mind in years. As they sang, it was almost as if it was happening again, and knowing the outcome, I could relax, laugh, and enjoy it. Life is often surprising, and we never know what tomorrow will bring. Today it filled my heart with happiness and thankfulness. Our friends and family were enjoying the fun and celebrating our fiftieth anniversary with us!

CHAPTER 18
GOING TO HORSE SHOWS

Jeannene called at the last minute and told us the owner of the barn where she boarded her horse was coming to a show near us. Being a horse trainer, the owner wanted to see our horses. We had been to her barn numerous times to see Jeannene ride, and she had heard lots of stories about Marlon and Dan.

Not having planned on going, we weren't prepared. We rushed to get our horses ready, loaded them in the horse trailer, and drove to the showgrounds. Arriving the day of the show, we had to take a dark, inside stall. Marlon couldn't comprehend why he was in such a place, and he was determined not to stay there. He fretted and made jerky movements that I had to dodge, finally employing his most effective kicks to the wall, which told me how unhappy he was. The look on his face showed how indignant he was that I had considered such a stall. I put him in a friend's stall while it was empty, took him outside, and walked him around, but nothing helped. There wasn't any way to improve his mood.

After only a few hours on the showgrounds, and after briefly meeting Jeannene's trainer, we loaded our horses into the trailer and went home. Dan hadn't complained at all.

"Next time, you'll drive through the showgrounds, wave at me, and keep on going," Cathy remarked.

Sometimes we became so exhausted toward the end of the three-day horse shows that we'd leave early. I usually got my early-morning classes in on the last day. Sometimes Bob got tired of waiting for his late-afternoon classes, and we'd leave. This prompted the smart remark from Cathy when we stayed only part of a day due to Marlon's behavior.

Quarter horse shows were held at this arena several times during the year. We went if possible. We could go home at night and leave the horses with Cathy, since she stayed on the showgrounds. Shows can be put on by anyone in conjunction with the Quarter Horse Association. A middle-aged couple, Martha and Jessie, sponsored these shows. They encouraged me with my unruly horse, who was prone to having meltdowns in the ring. Martha ran the shows, and Jessie took care of the barn and grounds. Diane had shown Marlon here for their first show-ring experience, and it was an apocalyptic event. No one who witnesses such behavior ever forgets it, and the horse's reputation is forever established. Much later, when I began entering classes on Marlon, we were watched, and everyone wondered what he would do that day.

After many mishaps and many rides, we began earning a place in the classes. To me it represented winning, even if there were only five people in the class and I got fifth. At least we weren't disqualified; if you were, you didn't get your name called. Time marched by as I entered class after class.

Then one day, Marlon and I clicked in my novice Hunter Under Saddle class. "Fifth place, Nell Parsons riding Sonny Corleone" came over the intercom. A respectable placing with eight to ten in the class. I was so happy and could not stop smiling. To me it meant my horse's behavior had greatly improved, and my riding was better. After we left the arena, I dismounted and hugged Marlon over and over and told him, "I am so proud of you." When I looked up, Jessie was watching. When we made eye contact, he laughed and gave me a big thumbs-up. Jessie and Martha had witnessed so much of Marlon's unruly behavior. I appreciated the encouragement from someone who had been there from the beginning.

Marlon did become more affectionate as he began to trust me. It took so long and went so slowly, until it almost seemed like it wasn't happening. I had to constantly adjust and to realize things were getting better between us. In the beginning, I thought he was angry every time he kicked his wall. I noticed later that they ranged from angry kicks to light taps. He used the softer taps to get my attention. Sometimes he wanted more of the good alfalfa hay instead of the grass hay. Other times he wanted more grain. After each kick, he would give me his inquisitive look to see my reaction. He sent different messages relating to what was happening, almost like Morse code. I quickly learned to give a response whether I agreed with him or not. Earlier I hadn't communicated to him when I understood. He needed me to let him know. When he got his way, he was a happy horse. When things didn't go his way, a big fight would occur.

Dan and Marlon were given carrots when their day's work was done. As I finished my chores so I could get the carrots, Marlon would rush me. He didn't know what a carrot was when I got him, but now carrot time was pure happiness.

After the carrots were gone, I'd hold my hands out to show Marlon I didn't have any more. He'd smell my hands and my pockets over and over and show his displeasure by raising his kicking leg to tap the wall. When I caught him, I'd say, "*No*," and he'd put his foot down. As I turned around to leave the stall, I'd hear the delayed tap. Glancing back at him, I'd see his mischievous look, and it was priceless. He was thrilled he'd told me how unhappy he was and that I had understood. I couldn't help laughing.

When he was relaxed and happy, he lovingly reached over and nudged me with his nose the entire time we were together. To me, this said everything. His beautiful dark eyes were soft and smiling. This melted my heart, and I temporarily forgot his problems. If I had concerns other than my Marlon problems, they were laid aside the minute he came into my sight. When he was happy, I could be happy. If he had a problem, nothing else entered my mind. It took all my concentration to deal with him. I lectured myself to feel hope and happiness on the bad days. He was more expressive with his feelings when he was away from home, especially at horse shows. I think he worried that I could go home without him. On his ugly days, it was total happiness if I could drive away and let Cathy deal with him.

He never quit telling me which way we should go if I was leading him or riding him. We would be headed in one direction, and Marlon would change course. "OK, this way is as good as the other one," I would say.

Cathy ignored this as long as she could. Then she started correcting me for letting Marlon make the decisions. There was so much for her to teach me that she'd let this slide in the past. I had needed to concentrate on more important problems—sitting still in the saddle, keeping my legs back, posting in sync with the horse, and so on.

"You must tell him what to do," she said firmly.

"How do I get him to listen to me?" I'd ask when he was misbehaving.

"He will listen to you once you have established authority."

When I asked her how to do this, the answer was always the same: "In time." Much later, she told me I was now riding with more authority.

My first reaction was, "Really? What am I doing different?" It had been such a slow process, I couldn't tell any difference.

"You are now saying, 'OK, Marlon, this is what we are going to do,' and then you do it. Look how happy he is that you are in charge. He will forever try to tell you what you should do. It is his personality, but most of the time he'll be happy to obey."

What relief to hear Cathy say I had taken another step forward.

Often when we were having lessons at the barn or at horse shows, other clients of Cathy's would be there. Going into the center of the ring to stand for a few minutes to rest or to let the other riders have room to execute a new maneuver was common. This was Marlon's dream and what he strived for. When riding, I had to be alert, or he would drift toward the center of the ring. This couldn't be tolerated because he'd do it in a class. Judges spot the slightest drift and will mark you down for losing control of your horse. After the riding lesson, we'd gather in the center of the ring to talk. Marlon was happy until he decided we had stood long enough, and then he'd start telling me it was time to go. He'd reach around and nip at my boot and start moving toward the barn. If I was on the ground, he'd bop me with his head to encourage me to go. Finally, I'd tell him, "You win!" Then we'd head for the barn, especially if he'd been a good boy that day. He thought carrot time was next, but if we were at a show, he had to complete all his classes first.

The majority of the barns on the showgrounds are large, with a wide hallway running down the center and stalls on each side. They usually have center entrances and one at each end of the building. The hallways can be packed dirt or concrete. Saddle racks with saddles, bridles hanging on doors, tack boxes, feed sacks with grain, and horse blankets for warmth or to cool a horse down after a strenuous workout can be viewed down the long hallway. There are chairs, fans or small heaters, show clothes hanging everywhere, and coolers, all for the use and comfort of the contestants. Everything is neat and organized upon arrival. As time goes by, the hallways become messy as the riders grow tired, disappointed, or decide to visit with friends instead of competing. It was interesting to try and guess each person's agenda: ride a little and have fun, dead serious on winning, a healthy attitude about doing your best and accepting the outcome, or get angry, pack up, and head for home—but not before blaming the poor judge.

Learning early that a ride without incident was a successful day for me and Marlon, I celebrated those days. If we calmly walked out of the show ring, it lessened my disappointment, even if we hadn't done well. I also shared some of the blame for our troubles, often entering shows for learning and riding experiences before being ready. Even with my failures, I still felt pride that I had tried and lived to try again. On the days that Marlon made a fool out of me, I still felt joy that he was mine.

On the showgrounds, Marlon liked to have me in sight. He constantly asked me what we were going to do. He enjoyed the time we spent together, but he knew the normal routine and had a keen sense for what came next. Schedules were different at each show. The lineup of classes, the number of classes, and the entrances varied for each show. Also, some judges were slower than others. It was difficult to convey to Marlon that each day was different and he must have patience. I worked hard to help him understand and gave him clues when it was close to time for his participation. Saddling him in a timely manner before our class was a necessity. Sometimes, if he was saddled too early, he'd get nervous and continuously ask me, "Isn't it time to go?" He'd swing his head and shoulder around to give me a firm bump, then stare at me with his big, inquisitive eyes and his ears straight up, listening for an answer.

He had me trained to do everything I could to help him calmly wait. I talked to him in a soft voice and petted him. If the wait was long, he got extra hay. This didn't always help. Sometimes he'd be too tense to eat.

Cathy felt I was too willing to do whatever Marlon wanted. He had misbehaved outrageously at the last show, so at this show, we started the process early. Cathy rode and warmed him up, hoping to emphasize that today he must behave. I rode next and had a good ride.

Returning to the barn, I asked Cathy, "Since it is a long time before my class, shouldn't Marlon be unsaddled so he can relax and possibly urinate?"

"No! He can certainly pee with a loose saddle on, and besides, he needs to learn to cooperate if he wants an easy life," she said.

"I sure hope Marlon understands your message," I told her. He did behave better, but it wasn't clear whether it was luck or if Cathy had gotten her point across.

I didn't enter the halter classes. Marlon was an elegant and graceful horse, shiny, dark bay in color, long-legged, and tall, coming from his thoroughbred ancestry. But he didn't have the muscular body shape for the quarter horse halter classes. Watching Bob and Dan in this class, I knew Marlon wasn't a halter horse. Dan did well and won enough halter points for them to qualify for an all-round amateur championship. They had to have points in a variety of classes, such as western pleasure, horsemanship, trail, and western riding, besides the halter points. For every five people in a class, one point is awarded to the winner. In large classes, you can earn lots of points if you win. With only ten people in the class, you still earn one point for second place. Dan was on the stubborn side and quick to know whom he could take advantage of; otherwise, he was a steady, dependable horse. He didn't see scary things around every corner, nor did he have flashbacks to haunt him. He'd had a dependable, steady life and was well adjusted. What a wonderful asset for anyone, animal or person! Bob was more high strung than Dan. They clashed at times, but no one could compete with Marlon and me for the most volatile duo. Even on his calm, happy days, it didn't take much to set him off. If I didn't find the cause, falls would often be the norm for me.

When Bob and I traveled to see our children or take trips, we boarded the horses with Cathy. If we were stateside, I often called to check on them. One time I called and talked to the barn manager, and she said, "The horses are fine." Then she added, "Does Marlon have anything wrong with his back ankles? He keeps reaching back and scratching them, which is hard for him to do. Each time he sees me, he does that. I keep examining them, and I can't find anything wrong. Maybe he just wants more attention with you gone."

"Sounds like fleas to me. Spray him with a flea-killing insecticide, especially his legs and ankles," I told her. As she was talking, I remembered that a mama dog with a bunch of puppies had recently been put in the small indoor arena behind Marlon's stall.

I called the next day, and Marlon was no longer scratching his ankles. Having worked with horses for many years, The barn manager was amazed at his ability to get his message across, and how determined he had been for her to understand.

"Well, I knew he was a smart one," she said. "Bless his heart, if he could talk, he sure would have a lot to tell us."

"Amen to that!"

One of my falls occurred in that small indoor riding ring before the mama dog and her puppies took up residence there. The weather wasn't good, so Cathy was giving me a lesson in the ring. I could hear noises nearby from the barn full of horses, and this made me nervous, but it was my only choice on this rainy day. Circling the ring, Marlon heard the different sounds and reacted by giving big swerves toward the center of the ring. Cathy was called into the barn area and left me riding. As I rode, more loud noises came and went. Marlon was erratically dancing and prancing when a critter ran across the ceiling above us. He jumped or tried to bolt, and I toppled off. There was plenty of sawdust covering the floor of the ring, so the fall wasn't bad. Cathy rushed back in to check on us. She and I were trying to figure out what had caused the loud noise when a cat crawled through a hole in the ceiling. I nervously got back on my scared horse. Marlon kept listening for other loud noises.

Ariel, Cathy's daughter, was a young beginning rider, and she rushed in to see what had happened. After explaining, Cathy told her, "Nell is a tough cowgirl; she didn't let a fall stop her. She got back on her horse and rode."

I had heard this expression from Cathy before. I knew she used any situation to teach her clients and especially her young daughter to never give up.

CHAPTER 19
TRAINING MAGGIE

With all the animals we accumulated over the years, I was their main caretaker, and I thought I could train any animal to some degree. I had experienced lots of disasters, but I learned and made progress, although I often had to settle for less than I wanted to accomplish.

Thinking back to my training days with our other animals, I went over each success and failure in detail to see if any could be applied to Marlon. Floating to the top of my list was the day Bob said to me, "I have a surprise for you!"

Soon I was picking a black-and-tan Doberman puppy from a squirming bunch of littermates. I had mentioned several times how neat, dignified, and alert these dogs were. Maggie, as she came to be known, came home with me and became my ever-loving companion and guard dog, whether I wanted or needed one. Her registered name was Magnolia Gail, and she was ten weeks old and adorable. She grew fast and furious and soon was a ball of energy in need of full-time exercise. We didn't have a fenced yard, but it became an urgent need.

When Maggie arrived, we were still working on our barn and fencing the pastures. Diane's horse was still with us and had to be cared for, as she was busy working and going to school. I did most of the outside chores and took Maggie with me, hoping she'd run around and expend her excess energy. Things were going well until a country dog that ran through our property chasing his owner's truck noticed Maggie. He raced around her and enticed her to follow him. Bursting with excitement, she ran after him, and I was powerless to stop her. I discussed this with the neighbor.

"So sorry. I can't do anything about my dog running through your property," he emphatically said.

I was angry! How dare he? Complaining to Bob released some of my anger. Growing up in the country, I knew the rules. The choices were to adjust to the problem or solve it by shooting the dog, getting rid of the body, and never admitting to anything. I couldn't shoot the poor innocent dog, so I began the task of teaching Maggie to come when called, even when she had fun things to do.

Correcting Maggie when she was caught didn't help. I practiced running faster, but there wasn't any catching this quick-as-lighting dog. She thought we were playing a game, and she was the winner each time. I came up with a new idea and enlisted Bob's help in trying out a silent whistle. The advertisement stated it emitted high-frequency sounds that only the dog could hear, and the dog would naturally respond to it. It sounded too good to be true. We were reluctant to try, yet something had to work. Bob and I took Maggie into the front yard and let her loose. She smelled and sniffed close by for a few minutes, then started wandering around our large front yard. Soon she was leaving our property. Bob blew the whistle. After all, he had been waiting and was ready. We thought to ourselves, "Hey, maybe she didn't hear it." He blew again with all his might. He blew again and again to no avail! We called, yelled, and chased her. We finally grabbed her when she paused to smell a dead critter. We felt ridiculous and laughed at ourselves.

"Can you believe we actually thought this might work?" I asked.

Growing by leaps and bounds, Maggie explored farther outside her boundaries. Soon she was chasing the cows and their babies on the farm next to ours. This was serious! A large ravine ran lengthwise through the pasture, and the babies were falling into it as Maggie chased the herd. One day, I dropped her leash, and instantly she was gone. Then I heard a loud, thundering sound coming from next door. Running over to the gate, I arrived in time to see the cows and babies pass by with Maggie in hot pursuit, barking as she ran. I only caught her when she was exhausted. I discussed this with Bob.

"Let's try attaching a heavy metal chain to Maggie's collar. She can drag it along and get some exercise while we work," he suggested.

The next day he brought home a heavy large-link chain about three times Maggie's length. We went to the barn, and Bob attached it to her collar and let her loose. She stood a minute, then walked off, dragging the chain and holding her neck and body sideways.

"Look how pitiful she is," we said.

We felt sorry for her but hardened our hearts and decided we didn't have a choice. Getting busy, we didn't notice her sneaking off. When we realized that she was missing, we began calling and looking for her. Suddenly, we heard that all-too-familiar thundering sound coming from next door. We ran as if our lives were in danger, hollering at the top of our voices. Entering the pasture, we saw the herd coming down the hill toward the ravine at top speed. Maggie was at a full run, yapping to the top of her voice, with the chain flying wildly through the air, swinging back and forth behind her. As she passed by, Bob dove for the chain but missed and hit the hard ground. I took one section of the pasture and Bob another; we joined the chase as they raced through our area. Finally, Bob latched onto the chain with both hands. Maggie dragged him a few feet as the chain slipped through his hands, taking skin with it. Now he was wounded, we were angry, and we still hadn't caught her. Maggie stopped running while dragging the heavy chain when she became too fatigued to continue. We were past exhausted, had to treat our wounds, and still hadn't solved our dog problem. There wasn't any laughing this time.

The current situation could not be accepted. I lived on several acres of rolling grassy pastures with few neighbors and couldn't let my rambunctious juvenile dog loose. Dreaming up ways to conquer Maggie occupied my mind night and day. We had tried everything we knew and had failed miserably.

By accident, Diane discovered that when she rode her horse, Maggie would lope alongside them. As long as Diane rode, Maggie would run with her, but the minute she stopped, Maggie would look all around, then streak for the cow pasture. One day, I told Diane to go ride and take Maggie with her. I put on my old farm clothes and a big floppy hat, picked up my broom, and sneaked through a front gate into the pasture next door. I hid behind a huge tree several feet beyond the five-strand barbwire fence line. Diane rode as normal, then stopped and called for Maggie to come. Instead of coming, Maggie headed for the fence line. Just as she had her head and front feet on my side of the low fence, I whammed the tree with the broom and said in a low, gruff voice, "*No,* you get back there!" I kept saying it over and over as I banged the ground and walked toward her. She froze halfway under the fence with a terrified look. As I got closer, she cleared the fence and ran in the opposite direction, looking for a place to go back under. Shooting though the fence and running back to Diane, she looked back at me in disbelief.

"You even scared me!" Diane finally said, after laughing so hard she could barely talk.

Maggie followed Diane to the barn and stayed close to her.

The next day I let Maggie loose and kept a close eye on her. She played, ran around our pasture, and did not go next door to chase the cows. She never chased them again. I had caught her in the act, scared her, and won that battle.

The owner of the farm had bet me Maggie couldn't be broken from chasing the cows or his truck as it drove back and forth every day. I applied the same technique to stopping her from chasing the truck, but it took longer, and I had to be more creative. After this, Maggie was entered in dog obedience classes. I needed to work on her doing what she was told without having to catch her in the act and scaring her half to death. The classes trained me as well as her. Maggie pushed me to the limit to get her under control. Bob did his fair share in helping me. Our struggles paid off, and she became a well-behaved and wonderful country dog.

Reviewing this and searching for any clues that could be applied to Marlon, I realized his problems were different. I still felt that love, consistency, and discipline when he misbehaved were all that could be done. He could not be punished for flashbacks. I hoped to convince him that I would never be unfair and that he would always know what was being asked of him.

CHAPTER 20
MARLON FALLS TO HIS KNEES

"What a great ride! Marlon, you did your part to perfection! Can't quit now!" I said as I finished riding one warm July morning.

I stopped in place for a few moments, then asked Marlon to back. He did it like a pro! Again I asked him to walk and then eased him into a steady canter. We'd gone halfway around the pasture when he hit uneven ground or a small hole. He fell down on both knees. I'd seen a horse do this in a thoroughbred race recently. Marlon put his head on the ground to push himself up. I could only help by sitting still in the saddle and not hindering him. In the many years we had been together, he had stumbled but never fallen. It scared him and concerned me as to what had caused the fall. I quickly dismounted and looked him over. I couldn't find any problem, so I got back on, and we walked around to see if there was any difference. He seemed fine. The next day, he was sore and walked funny. I checked him again and discovered he had a loose shoe.

For any kind of mishap that Marlon and I encountered, my human riding buddy was present and gave his opinion on my riding progress. One thing about Bob: he was witty, yet he told it like it was, good or bad. That day, I earned high praise for keeping my balance as Marlon fell and struggled to get back up.

Some mornings, we decided to take the day off from riding. One of us would give the excuse that we were sore and tired.

"Think I'll saddle Dan up and ride a little," my dear husband would say a few hours later.

Of course, if Dan left the barn, my uptight horse must go. I'd have to readjust my thinking and my schedule for the day and ask Bob to wait for me.

Alabama has plenty of flawless horseback-riding days, with the sun shining bright, the temperature cool, and not a breath of wind. This was one of those days. I was so appreciative and hoped nothing would happen to dampen my enthusiasm.

When we finished our workout, Bob and I walked the horses back to the barn. As we dismounted and let the reins go slack, Marlon's bridle fell out of his mouth. The little roller in the center of the bit fell apart. I was stunned at first, then thankful that it hadn't happened while riding, or at least I hadn't been aware of having zero control.

Until a replacement bit could be found, I rode with his show bridle. It is a snaffle bit that lets the horse move forward freely and smoothly, as he needs to do in the show ring. Amazing feeling! In the past, I didn't ride out in the open with the snaffle because the roller bit gave me more control. Now I didn't have a choice. Sometimes this is what it takes to force you to move forward.

Marlon's internal clock was always on full alert. It was fine-tuned to the minute. From day one, we had gone to the extreme to keep the horses on a set schedule because of him. Besides feeding in a timely manner, we strove to keep turnout and riding times the same. On a very hot day, we decided to ride after dinner, when it was cooler. We fed the grain to the horses early. They eat their hay for hours, so we held it until later.

Marlon didn't like me coming into his stall at this hour with my grooming box. He kept telling me it wasn't time to clean his feet, brush the sawdust off of him where he had been rolling, and prepare him to be saddled. He protested by swinging his body side to side, staring at me, and asking me why I was there. It was a long, aggravating process to get him ready for the saddle. Things went downhill fast when he saw me approaching him with the saddle. I began to wonder what he would do when I got on to ride.

After putting the saddle on, I walked him outside the stall and put my stool alongside him. He kept moving just enough to keep me from getting on. I'd climb down, move the stool, and step up again. Over and over it went. Finally, I had to give him several open-handed smacks before he would stand still. The smack on his neck didn't hurt him but let him know I would keep on until he behaved.

As we rode, he kept his eye on the barn and drifted in that direction. One of his favorite tricks was to slowly lumber along going away from the barn and speed up going toward it. His message was that he would tolerate me riding at this odd hour if we'd hurry up and finish. I decided I'd ride in the heat next time.

Back at the barn, Marlon was still rushing me to unsaddle, spray him off, and let him back in his stall. Once in his stall, he nickered and kicked the rubber-lined walls before I had time to give him his hay. I looked over at Dan as he stood quietly and uncomplaining while Bob took the big western saddle off him and completed his chores.

Bob laughed. "You chose him. Don't say a word," Bob said when he saw me watching them.

"True, but I can complain a little if it makes me feel better, right?"

CHAPTER 21
MY PERFECT RIDE

Columbia, Tennessee, was a favorite place to show our horses. The showgrounds were located in a beautiful park, the people were friendly, the food was good, and it was very relaxing. Portable stalls were brought in to accommodate the crowd for quarter horse shows. The rings were outside, and weather could be a problem, but when the weather cooperated, it was great. Another advantage: it was only an hour away from home.

Special memories were made at one of these shows. Marlon and I had a perfect ride in an amateur Hunter Under Saddle class. To me, it was my only flawless, perfect ride in a show ring. I had lots of great rides, but seconds of change in Marlon's flow, which I could feel, or a moment of hesitation by me at some point, would make it slightly less than perfect. The day didn't start out well for me. The ring was at the bottom of a sloping hill. It wasn't steep, but it worked best for me to carry my mounting stool and walk Marlon down to the ring. As we arrived at a good area to get on my horse and leave my stool in an out-of-the-way place, a group of cowboys was standing nearby. They stopped talking, tipped their western hats, and smiled, and all I heard was, "Ma'am."

Marlon was wearing an English saddle; I was an older lady decked out in knee-high boots, fitted riding pants, hunt jacket, and safety hard hat. I nervously smiled while positioning my stool at the best place for my struggle to mount my tall thoroughbred quarter horse. In my side vision, I could see them intensely watching, but I was committed to getting on my horse here. Giving it all I had, I managed to lurch up into the saddle with one try. I settled myself down, gathered my reins, and walked past the cowboys. They gave me amused smiles and again tipped their hats.

I warmed up in the ring. The English classes were next. It was chaotic with so many riders. I wasn't feeling too confident, but I kept Marlon safe and out of everyone's way. After a short warm-up, the ring was cleared of riders, and the ground was leveled. Time for classes to start, and my novice Hunter Under Saddle class was coming up soon. Marlon and I walked around until my class was called. Moving around helped to keep our nerves calm.

Entering the ring, I noticed there were lots of riders. Having a larger number of participants in a class makes it easier to be in someone's way or for something to interfere with your ride. This put me on edge, but my ride went well until Marlon broke his stride in front of the judge. It happened so fast, I couldn't tell if he had a reason or not. I kept wondering if I should have done more to help him. Riding is a partnership between horse and rider, and the rider is supposed to be in control. When the class ended, I placed low, although my name was called. So disappointing! We had done well except for one little glitch at the wrong moment.

Cathy and her other clients were discussing my ride when a friend suggested I enter the amateur class. It was coming up next and was an advanced class. I was just a novice, without much experience.

I was leery, but the others joined in, saying, "What do you have to lose?"

Bob raced up the hill and entered me in the amateur class. When it was time for the class, everyone entered the ring, the gate was closed, and the announcer called for everyone to move to the rail and pick up the trot. Most of these amateur riders were experienced and often won. Everyone scattered, finding their place on the rail, and the judging began. Immediately, I could feel Marlon was moving at a smooth, steady trot; we were flowing around the ring like pros. He changed gaits in his past winning way. Cruising by friends scattered throughout the bleachers, I could see them in my peripheral vision. They kept encouraging me in low voices so as not to startle my horse. It felt like Marlon and I were floating on air, round and round the ring. It was magical! To me, it was perfect.

After we completed three gaits in each direction, the judge walked to the center of the ring, signaling for the riders to line up in front of him. Marlon finished the class with a smart, unhesitant back as the judge walked down the line for final judging. I relaxed with a sigh of relief and waited for the class placements. When the winners were announced, we came in at second place. A loud cheer erupted from friends and other observers. Today everyone agreed with the judge. I was ecstatic to place in this class and have my one perfect ride. So happy that Marlon and I were in complete harmony this day!

CHAPTER 22
UNUSUAL THINGS SCARE MARLON

Another lovely horseback-riding day was before us, with the angle of the sun letting us know fall had arrived. We went to the barn to ride, have a lesson, and see our buddies.

When I arrived at the barn, the first question I asked Cathy was, "Should I lunge Marlon today?" He usually let me know if he was feeling nervous, antsy, or angry. Sometimes he didn't realize how much energy he had until we started riding.

"It never hurts to give him a chance to kick up his heels," Cathy answered.

Marlon was relaxing and trusting me more as the years went by, although it was still a struggle to read him correctly. He was so lightning quick when leaping around, I needed all advantages that I could get.

I lunged Marlon, and he wasn't enthusiastic about it. I had to make him go through the three gaits in each direction. Afterward, I groomed him, put the saddle on, and mounted to ride. He was still on the lazy side, so I firmly encouraged him to get into his smooth gaits. We'd been riding for a short time when a goose in the pasture next to our riding area came parading by on its way to the pond. This was normal, and I wasn't concerned. I noticed about the same time as Marlon that three little ducks were following the goose in a perfect line. This put a whole new spin on the situation for Marlon. He began moving around erratically with his head up in the air, his ears forward on high alert, continuing to stare at the parade.

"Dismount now!" Cathy yelled.

It was such a surprise to be told to get off my horse. I'd seldom dismounted in any riding situation. I was usually encouraged to ride it through. The day before, when Cathy was riding Marlon, the parade had passed by, and he'd gone crazy. We couldn't figure out why adding the three ducklings made such a scary scene for him. After they were out of sight, I put him back on the lunge line. He ran and ran. As he raced around, he kept looking for the goose and followers. After his energetic workout, I climbed back into the saddle, and we had an excellent ride. He did keep a sharp eye out for the parade, though, and I helped him watch.

Returning from a trip to Peachtree City to see Julianne and Jeannene play soccer, we were crossing Sand Mountain and nearing the stables when we noticed the sky had the greenish hue that accompanies tornadoes. We turned on the radio and realized we were driving into the storm's predicted path. Stopping at the barn, we saw the doors were closed, and not a single animal could be seen. As we stepped through the double doors, all the barn dogs and cats met us. The thirty-plus horses called to us and told us how scared they were. They knew the storm was coming. We couldn't move without stepping on or kicking a dog or cat; they were under our feet constantly. The horses wanted reassurance that they'd be safe. Dan had a calm personality, and he wasn't showing signs of nervousness. Bob put the halter on him, and they walked up and down the hallways with the dogs and cats following. It helped to calm the horses. Marlon was so upset he had his head in the back corner of his stall and was kicking the wall constantly. Talking to him at a safe distance and saying, "I am with you" was all I could do to comfort him. Bob, Dan, and their entourage walked the halls until the storm passed. We stayed on afterward to make sure the animals settled down.

Another big nightmare for Marlon was hearing and seeing fireworks being shot into the air. Most years, he was home over the Fourth of July, and in the county, you are allowed to shoot fireworks on your property. The neighbors always had a party with plenty of fireworks zooming into the air, making a spectacular scene. They would notify us when they planned to start, and we'd prepare as best we could. The radio and fans in the barn were turned up to the highest level, and the barn doors were closed. I didn't want Marlon to hear and see the hissing fireworks exploding into big balls of fire. This helped some, but I dreaded this celebration for Marlon from year to year.

One spring, when the horses came home, they discovered that our neighbor had purchased two llamas, and they were in the pasture next to Marlon's. The horses backed out of the horse trailer and stood a moment to let the home atmosphere sink in. Marlon couldn't wait to get in his pasture and celebrate. Movement in the previously vacant pasture caught their attention. Invaders like these had never been seen before. The horses stood staring, shaking all over, with their hearts beating wildly. Every now and then they'd run a few steps, then stop and stare. For weeks, Marlon couldn't go into his pasture that shared a fence line with the llamas. He had to take turns with Dan in his pasture, which was farther away. It was a long adjustment period for Marlon. Even calm Dan needed time to come to terms with this. There wasn't any hiding from the llamas, as they never left their pasture. We couldn't ride for weeks. Bob could ride Dan long before I'd even think about getting on Marlon. Months later they became friends with the llamas, whose names were Grady and Packo.

My friend from Virginia came to visit. She asked Bob and me to dress in our show clothes and demonstrate our riding abilities. It wasn't much of a stretch for Bob to get ready for riding western, but she wanted me in full English attire.

If I was to be dressed in my finest, Marlon had to look his best too. First I groomed him to perfection and then saddled him in his English tack. I explained to her what was involved to get a horse ready for the show ring and the work required to keep your tack looking good and in working condition.

Next I put on my full English outfit, topped off with my safety hard hat, and climbed into the saddle. I made my entrance from the barn into the riding area in the pasture. Bob and Dan were already riding and showing her what they could do.

Earlier I had put a chair inside the gate to the pasture and positioned it where my friend would be a safe distance away but would have a panoramic view of us riding. Seeing me, she was impressed and thought I'd fit in with the younger competitors at the shows. I started doing my warm-up routine in front of her, getting Marlon ready for our little exhibition. I was turning him in a tight circle when a loud hissing noise came from the farm down below. It was so sudden and so loud, it absolutely terrified me and Marlon. He just lost it and started doing anything he could. I was saved from falling off as I was turning him in a tight circle and froze, holding him in that position. He couldn't bolt forward, and it hindered his ability to jump, whirl, and dance.

My friend leaped up, grabbed her chair, flew out the gate, and closed it behind her.

"Get off that horse," Bob yelled.

There was no way to dismount as we continued to whirl around and around. After several minutes the noise stopped, but before I could settle back into the saddle, it abruptly started again. We had another round of the same. Suddenly, the noise paused again.

"Whoa!" Bob screamed several times as loud as he could.

It startled Marlon, and he paused for a few seconds. I quickly dismounted. Marlon was very upset. We didn't know what the sound was or if it would start again. My riding exhibition for my friend ended before it began.

I unsaddled Marlon, put him up, took off my show clothes, and tried to figure out what had happened. Later our neighbor told us he had been removing rust from some metal farm equipment and had turned the air compressor on the rust spots at full force. The result had been an unexpected high, shrill sound. He promised he would look, and if we were riding when he was going to do something with a loud noise, he'd warn us.

Marlon continued to listen for and expect the noise to start again for weeks. Each time we went out to ride, I helped him check the landscape for anything unusual.

Later we laughed about my almost ride, but it had not been funny at the time.

CHAPTER 23
MY LAST WILD RIDE

I didn't intend for this to be my last ride in competition, but circumstances made it so. It wasn't a good way for me to end my show career, yet it was entirely appropriate for my partner, Marlon. For over seventeen years, he had tried to tell me what to do. In fact, he did everything he knew to convince me he knew best. As a result, our last ride in competition was the wildest ever.

The previous fall, Bob and I had gone to a horse show on the spur of the moment. We hadn't prepared and didn't have the edge needed for competition. In my first Hunter Under Saddle class, as Marlon and I passed the announcer's stand, the microphone made several high-pitched screeches. It startled Marlon, and he gave a big swerve away from the sound, throwing us off the rail and into the paths of the other riders, causing all kinds of problems.

Before I corrected the situation, the judge motioned for us to come to the center of the ring and stand until the class was over. It had just begun. Marlon was thrilled he had made it to the center of the ring in record time, signaling that his participation in the class was finished. This was embarrassing for me, as I was responsible for our predicament. I hadn't reacted quickly enough.

The following spring was our first time returning to the horse show arena. There wasn't any doubt in my mind that Marlon would remember our previous experience. Still, all these months later, I didn't expect him to insist we go to the center of the ring the moment the class started. It was an all-out war between us from the beginning to the end.

In my first class, the judge called for everyone to enter the arena at a trot. Marlon trotted and cantered at the same time, a few steps of one before switching to the other. I pulled back on the reins to slow him down, but he kicked out with his back legs and bucked, throwing his rear end up into the air as high as possible. Every few steps he'd give a high squeal and lunge toward the center of the ring. He was fighting me with everything he had, making all the moves that had gotten him to the center in record time last fall.

With his wild and erratic moves, I was terrified, knowing I could sail through the air and hit the ground at any moment. This was tempered by my struggle to manage a determined horse. I did move Marlon off the rail, out of the path of the other riders. When the canter was called for, Marlon understood the announcer and instantly leaped out at an uncontrolled gallop.

"This must be like riding a bucking bronco," I thought. I panicked. It was all I could do to hold him back to a very fast canter. Passing Cathy, who was standing ringside, I glanced at her for words of wisdom. She just shrugged and didn't say anything. We made many circles around the arena, and I looked each time for instructions. There were none.

Toward the end of the class, when the judge asked us to come to the center and line up, I apologized to him and said, "This is his first class of the season. He's never been so bad."

"I kept a close eye on you to make sure you weren't interrupting the class. You weren't disqualified, because you did all three gaits in each direction," he responded.

Walking out of the arena, I found it hard to believe what had just happened. I was so relieved that Marlon hadn't tossed me across the arena and that we hadn't interfered with my fellow riders, but I was totally embarrassed. Once again, Marlon and I had put on a show for all spectators. Yet deep down I was excited, knowing this experience had been the ride of a lifetime. Approaching Cathy and her other clients, I heard them talking about my ride, and I heard Cathy say, "She is tough!"

I knew she was saying this for the benefit of her new clients.

"I sought advice each time I passed by!" I exclaimed.

"You were handling the situation, and there wasn't anything I could add," she said.

At that moment, the adrenaline was flowing too much for me to realize this was the first time she hadn't given me advice. I couldn't help thinking our escapades had finally become too much for her.

We decided Cathy would ride Marlon in the open Hunter Under Saddle class. She needed to convince Marlon that he couldn't go stand in the center until the class was over. He battled her to get there, but he'd expended most of his energy and didn't fight as hard.

My emotions ran the gamut from one extreme to the other. The thrill of a fast ride had turned to terror when he started bucking and kicking, and then worry had set in that I'd fall off in the arena and make more of a spectacle of myself. Then anger at Marlon was thrown in the mix for doing this to me after all our good years. The last overwhelming emotion was happiness that I had advanced in my riding ability enough to handle and survive all that Marlon had thrown at me this day.

Later I was standing ringside when a couple of cowboys walked up to watch Cathy's class. As Marlon and Cathy passed by, they made the connection that I was the one who'd had the wild ride. They stared at me with startled expressions for a moment, then burst out laughing and couldn't stop. They tried to talk but could only utter a word here and there.

Between bouts of laughter, they finally managed to say, "Hey, you hung in there! That was some wild ride! We couldn't believe your horse never gave up! Boy, you did great!"

By then, I was able to laugh and talk about my ride. They gave me high fives for surviving the ride and finishing the class. It was a humbling experience for me, as our ride was the main topic of conversation all day.

Sometimes I wished that I'd gone to additional shows just to see what Marlon would do, but I always ended up feeling our show career ended just as it should have. Everyone knew my and Marlon's long history, so it probably was a fitting last chapter. After that, but not because of it, Bob and I made the decision to give up horse showing. We were well into our seventies, and it was too much work getting us and the horses ready to show. My English classes were early in the morning, and Bob's western classes were late in the afternoon or evening, making for long and tiring days.

We usually left on a Thursday to travel to a three-day horse show. It took days to get the trailer loaded with hay and grain for the horses and all the equipment needed—the tack for warm-ups and the polished and shined show tack, plus the clothes needed for the classes. We had to have both English and western tack and show clothes. Small suitcases were packed to take to the motel for our short nights. Other necessities included tools for emergencies, medical supplies for accidents, and a large cooler for snacks and drinks. The pink scooter was added for running errands.

Most shows ran Friday, Saturday, and Sunday. If the Sunday shows were running late, we'd often have to leave before Bob's classes were finished. As the years rolled by, sometimes he was ready to leave by the time mine were over. We mainly showed in Alabama, Tennessee, and Georgia. In those early, exciting days when we were younger, we traveled to Mississippi, South Carolina, and Florida for horse events. After a long drive and many tiring days, it would take us days to recover, but by the time the next show rolled around, we were again ready and excited for the challenge. Finally, it became too much, and we retired from competition. We missed our horse friends, but we still rode with Cathy and often saw some of them at her place. When we were feeling nostalgic, we celebrated that we'd had this opportunity to pursue our hobby for years.

CHAPTER 24
BOB'S SURGERY AND THE HORSE COMMERCIAL, 2007

After retiring from competition, we continued to ride our horses three to five times a week, look after our five-acre mini farm, and take care of our pampered horses. For the last few years, Bob had been having some discomfort in his legs.

A vascular surgeon told him, "There is blockage in the arteries in your legs, but keep walking, riding, and working."

In September 2007, Bob woke up with excruciating pain in his left leg. Examining it, we noticed his foot was beginning to turn blue. This got our attention. We called the doctor, and he told Bob to come in immediately. After the examination, Bob was sent straight to the hospital and put on a blood thinner. It was late in the week, and surgery couldn't be scheduled until the following Monday morning. Bob spent several long days in the hospital waiting for surgery. I spent the time getting the chores done and preparing for the unknown.

The horses were still at home. I called Cathy to see if she could come and get them. I didn't have an extra day to hook up the truck and trailer and drive an hour each way to her barn. I needed to spend my time packing all the horses' necessities and loading everything into the trailer. With cool weather approaching and me not knowing how long they'd be gone, all their belongings had to be taken. It took a couple days for me to get ready for Cathy to come and pick up our loaded trailer and the horses.

The doctor estimated Bob's surgery would take three to four hours. Many prayers were said over the weekend as I worried and prepared for Monday morning. It helped being busy. Bob's surgery was scheduled first, and he was told he'd be transported to the surgical area by 4:00 a.m.

I arrived at the hospital before Bob left his room for the surgical ward. "I am so happy to see you!" he said. "I didn't believe you'd have time to get here this early with all that you had to do."

Besides seeing him, I wanted to know where they were transporting him so I'd know where to wait. It was a large hospital with several surgical areas. As it turned out, I got to spend time with him in the prep area. When he left for surgery, I joined friends in the waiting room.

They told me reports from the operating room would be sent out every thirty minutes. Friends came and went according to their schedules as the hours dragged by. Everything was going as planned until I realized surgery was running overtime. It was getting longer between reports. Becoming anxious, I checked with the receptionist for an update. It was another half hour before word was sent that they were finishing up and that the doctor would meet me shortly in the consulting room. The longer time for the surgery and lack of information had made me frantic. I immediately went to the consulting room at the back of the large waiting room and impatiently waited for the doctor. He arrived looking exhausted. The blockage had been worse than originally thought, and the artery from Bob's ankle up into his abdomen area had to be replaced.

"I took the vein out, turned it around, and put it back in to act as the artery. It has to be reversed so the cut-off valves in the vein will go in the right direction," the doctor explained.

Bob worked hard to recover and get out of the hospital. The surgeon was accommodating, coming by to see him at all hours, visiting with him, and explaining everything in detail. So when the surgeon asked him to make a commercial for the vascular surgery department at Huntsville Hospital, Bob agreed.

Later he said, "They caught me at a weak moment, and I agreed. I don't like seeing myself on television."

At home, my job was to keep Bob's incision clean. The first time the gauze was removed and I looked at the incision, I was shocked. Silently, I counted seventy-seven staples. Getting up and moving around was tough, but Bob was determined to get back to normal as quickly as possible. Dan was waiting.

Five weeks into Bob's recovery, the advertising agency that made commercials for Huntsville Hospital called Bob and wanted to know what his hobbies were. "Is it golf or hunting?" they asked.

He happily told them he rode and showed his quarter horse. They arranged to come out to our place in October to film the commercial. Bob couldn't ride, but he and Dan could reenact their halter class. He'd lead Dan around, showing off how handsome his horse looked.

Before the commercial could be made, the horses had to come home. Bob could go along for the ride and back me into the trailer for hookup. When the day arrived for the commercial to be made, a policeman who helped with the barn chores came to bathe Dan and get him ready for his TV debut.

"I want to bathe Marlon also," I told him.

"You want to sneak Marlon into the commercial," Bob and the policeman teased. If the opportunity arose, Marlon would be ready, but I couldn't own up to such a thought.

I groomed both horses fit for the show ring, polished and shined Bob's silver-loaded show halter, and made sure his western clothes were ready. Knowing that blue photographs well on television, I chose a pretty blue western shirt for him to wear.

After lunch, a van and several other vehicles arrived with assorted specialists and equipment to shoot the commercial. The vehicles drove through the gate into our pasture surrounding the barn. Workers got out of their vehicles, looked around, then sprawled on the grass.

"Perfect afternoon and lovely scenery for shooting a commercial," they kept saying. "The old silver metal barn with stark white trim and three bright-yellow doors across the front will photograph beautifully. It sits in front of a thick green grove of trees that will make a perfect background."

They acted excited and soon began instructing Bob on what they wanted him to say and do. When they were ready for Dan to join Bob, I went to the barn to get him. Marlon's stall was opened first, and he went trotting out into the pasture. He had to go first or he'd be extremely unhappy, and

I would pay the price with a messed-up stall. The rule of always being first still applied for Marlon, even on Dan's special day.

Marlon trotted circles around the people and equipment, trying to determine what was happening. As he ran toward them, they weren't sure whether to run or stand their ground. He stopped nearby where he could watch and started grazing.

"Oh, how shiny and beautiful he is," they exclaimed.

Next I put the halter on Dan and led him out to where the activity was. They were still going over the details with Bob—where he should lead Dan, and how far he should walk before stopping to speak. I sort of suggested the best background for the commercial; however, they didn't need or want my advice. The background had already been chosen, and after a few takes with Bob and Dan, it was finished. Some still shots were done with a different background for advertisements in magazines and daily newspapers.

"Usually when we do this, it is a hundred degrees in the shade, or bitter cold," the director said. "Today is perfect." The crew didn't want to leave.

The advertising company sent us a copy of the commercial, telling us when it would start running. Looking at the video, I saw Marlon wasn't grazing in the background as I had secretly hoped. The commercial ran on all the local channels, any hour of the day or night. You never knew when. Friends saw it and were surprised. It was very popular because of the gorgeous horse. Everyone loves to see a cowboy with his special horse. Sometimes when Bob wasn't around, I'd hear him talking, stop in my tracks, and then realize it was the commercial. Bob asked, "Can you imagine how strange that makes me feel?"

Months later, at a birthday celebration, I was across the room from Bob when a woman from out of town said, "Who is that man over there? I know him from somewhere."

All the people within earshot stopped talking and looked across the room at a large group of men and women. "That man and his horse are on television. I see them all the time," she said, making the connection before any of us could react.

After a short recuperation, Bob was back in the saddle and working hard to build up his strength. His buddy Dan played a big role in helping him to do this.

Marlon had a hard time forgiving me for what he considered neglect while Bob was recuperating. They were brought home for the commercial to be made. He was well taken care of, and the barn helper came regularly to assist me with the horses. Marlon wanted me to spend long periods of time with him, and if I couldn't, he was unhappy. His security blanket wasn't acting like one. He saw me every day, but he wanted more normal personal time.

CHAPTER 25
JEANNENE'S VISIT

When Bob's leg problems had surfaced, we canceled a scheduled trip to Germany, where we had lived from 1961 to 1964. Our apartment building sat on top of a mountain overlooking the old city of Trier. From our balcony, we could see the city, the Mosel River winding its way through the deep, narrow valley down below, and the vineyards on the steep sides of the mountains. We wondered how the grapes were picked from such precarious spots. The colorful scenery was beyond description, even with so much scarring left from the war. On top of the mountain on the other side of the river was a French garrison. We had joint military socials with them, and we all struggled to communicate. Thinking of their few words of English and our couple of words of French is still vivid in my mind after all these years. The joke was that the Americans only knew two French words, *oui* and *mademoiselle*, regardless of how much exposure we had to the language.

Bob and I were discussing the historical sites in and around Trier when I remembered asking Jeannene what she had seen as a young girl on a trip to Spain. "A lot of old stuff," she replied. We were ready to see familiar old stuff again.

I enjoyed repeating this to Jeannene when she came to visit one weekend. She attended Berry College, a couple of hours' drive from our house. She came to see us but mainly to ride Marlon. By now, the girls were excellent riders and preferred to ride him over Dan. If you sat still in the saddle and told Marlon in his language what you wanted him to do, he cooperated and was a pleasure to ride. Dan was stubborn and made them work hard to get him to do what they asked, and it wasn't as much fun.

Knowing how expensive it was for a college student to do laundry, I suggested she bring her laundry with her. She was embarrassed about the amount she had. "Oh no, Mama, I can do it later," she kept saying. Remembering my college days, when doing laundry took all my extra money, I ordered her to bring it.

Arriving on a Saturday morning (she had worked Friday night at her college job), we immediately sorted her dirty clothes and started washing them. After having a second breakfast with Jeannene, we went to the barn. It was a very windy day due to Hurricane Gustav. I started getting Marlon ready for her to ride. She stood back watching and kept asking if she could get me anything, finally deciding to take pictures of Marlon with her cell phone. He told her he wasn't happy about her walking around at a safe distance and flashing a light in his direction. Complaining and lightly threatening her, he wanted her to leave. When I finished grooming Marlon, I took him to the round pen and let him exercise, then led him into the pasture, holding him while Jeannene climbed into the saddle. Off they went, both looking comfortable with each other. I watched them ride in case I needed to give her advice on some peculiarity of Marlon's. She was an excellent rider. Marlon appreciated her for it and promptly forgot being aggravated at her. They had a long, fun ride with Bob and Dan.

I unsaddled Marlon, put him in his stall, and went to the house to fix lunch. Jeannene stayed to help Papa with the barn chores.

"Marlon is the only horse I have ever been afraid of on the ground," she told Bob. "I have loved riding him all these years, but he scares me."

I'd noticed she didn't offer to help me get Marlon ready to ride. Knowing she had ridden show horses and jumping horses for other people and had a horse of her own, it surprised me. I thought she had overcome her fear as she grew older and as Marlon became a teddy bear, compared to his early days.

One day, I received this note from Jeannene expressing all four granddaughters' feelings about Marlon and their journey of learning to ride him.

"All of us were intimidated by Marlon, Mama's big dark bay, who more often than not pinned his ears back every time we tried to approach him. He'd show his usual ugly face and throw in a few hard kicks, making it known that we were in his space or that we weren't feeding him quick enough. To us, he was dangerous, mean, and scared us beyond anything we had ever witnessed. But Mama saw him with a different set of lenses. She saw the hurt and neglect he had experienced in his past life. She was patient, loyal, and stuck with Marlon, earning his trust. Through the years and without even realizing it, Marlon quietly and unsuspectingly became the horse that all the granddaughters preferred to ride. He became calm, trusting, and willing. Mama's consistent nurturing and love had quite literally transformed him. What is most remarkable is that she helped him trust us and other people as well.

"This quote by Konrad Lorenz captures the journey of Marlon and Mama perfectly: 'The fidelity of a horse is a precious gift demanding no less binding moral responsibilities than the friendship of a human being. The bond with a horse is as lasting as the ties of the earth can ever be.' We are so thankful we got to witness this friendship."

Over lunch I told Jeannene about Bob going to the pasture to get Marlon. He had hooked the lead rope to Marlon's halter, and they'd walked side by side to the barn. Leading him into his stall, Bob reached up to unhook the lead rope and discovered it had never been hooked. Marlon had walked by him in the correct place as Bob held the unhooked lead line. We laughed at Papa and reminisced about how amazed we were at the changes in Marlon's behavior. It was such fun to tell Marlon stories with people who were there from the beginning.

Bob and I hooked up the trailer to the truck late one evening, preparing to take the horses to Cathy's barn the next day. Marlon kept standing in his doorway, looking out and calling me. I finally decided he wanted to go to the round pen and exercise. I went to the house to put on my work clothes, and Bob started mowing one of the pastures. Walking out to the barn, Bob yelled at me and said, "I forgot to give the horses their grass hay." I put the hay in their stalls, but instead of eating, Marlon kept calling and looking out his doorway. Getting his message, I took him to the round pen for exercise. He was ready. I just stood there while he galloped, scratched himself off, jumped, and twisted, and when he finished, he walked up and stood directly in front of me. It was still unbelievable to me that he could be so precise in the timing of what he wanted and needed.

Dealing with Dan was a totally different story for me. I never had the upper hand with him and always knew I didn't have what it took to get it. Any time I dealt with him, I had to make sure I had the advantage. Still, he could find ways to overpower me. Having to take such precautions to simply lead him out to pasture was such a pain.

Dan was a gorgeous, muscular American quarter horse, and he placed in halter classes, which are similar to beauty pageants for people. He'd never been mistreated, and he grew up secure and trusting. Bob got along with him, although Dan was stubborn, extremely mischievous, and full of himself. Early on, Dan learned it was easy to overpower me; as a result, he didn't respect my authority. He loved to play tricks on me because he could win. I didn't like being responsible for him in any way. Tying him up to clean his stall was necessary, or he would try to crush me against the wall. I barely escaped a few times without getting hurt. For my safety, it was imperative that I learn to protect myself and not let him win.

This was vividly demonstrated to me the day I volunteered to lunge Dan for Bob when he was recovering from leg surgery. We lunged our horses in the round pen, which was filled with sand for good traction. If the horses were feeling unusually energetic, they could safely run in small circles and kick up their heels. Bob put Dan's bell boots on so he wouldn't pull a shoe, as well as splint boots in case he accidentally kicked his ankles or lower legs. Last he put the chain underneath his chin to keep him from overpowering me on the way to the round pen.

Arriving, I led Dan into the round pen, closed the gate, and turned him loose. I had lunged Dan a few times over the years, and he remembered the times he'd won the power struggle. He was always looking for a chance to show me he was in charge. I had to be on high alert at all times. When I led him, the chain under his chin hung loose unless he decided to bolt. Knowing whether it was there, he behaved accordingly. Waiting for the right moment to make his move, he'd start a slow jog, and if I didn't instantly pick up on it, he'd be a step ahead and could overpower me. Delighted that he was in control, off he'd gallop, kicking up his heels in victory. Catching him was impossible until he had celebrated to his heart's content. We had no choice but to wait until he stopped running and started grazing. This could go on for a long time with a well-fed and energetic horse like Dan.

Obeying all my commands in the first direction, he was enjoying the freedom to run, kick, and squeal. He kept a close eye on me, and I certainly was watching him. When he thought I was getting ready for a reversal in direction, he suddenly turned and ran straight at me. It was such a short distance that there wasn't time to react and raise the whip for protection. In that fraction of a second, I didn't see my life flash before me, but I did envision him running over me with his metal shoes on. I knew a serious injury would be the result. Just as he got close enough to touch me, he swerved and ran past me. Relief swept over me, as well as disbelief at what he had done. He was so proud of himself. He galloped so fast around the small ring that he almost fell down in making the curves.

His arrogant "I showed you who's in control" look made his point very clear, and I got it. Being scared and shaken up, I ran to the side of the ten-foot-high round pen and climbed up and over in one second flat, literally flying through the air.

I surprised Bob when I came marching into the barn and told him what his horse had done. We went back to the round pen, and Bob suggested we go in together so we could teach Dan a lesson. He needed to know he couldn't charge me or anyone else. It sounded like a feasible idea, although I was skeptical. Bob gave Dan the cue, and he started trotting around the pen. Standing there watching Dan, I could still see the gleam in his eyes, and I decided I didn't want to teach him anything, even if I could. Racing to the side of the ring, I once again went up and over the top of the round pen. Bob would have to do the correcting and teaching of his horse.

Bob knew how scared I was, but he was amused at a seventy-plus-year-old lady going over the high fence not once but twice. He couldn't resist telling family and friends how I was learning to "fly without a parachute."

The horses still did unexpected things, even though they had been with us for years. Everything would go along smoothly, and we'd begin to think we could predict their behavior. Then some bizarre incident would happen that left us shaking our heads in disbelief.

Marlon had been with me for fifteen years, and he'd always gone into the same pastures. In the early years, he went into the farthest pasture, with the middle pasture separating him from Dan. After Dan stopped challenging Marlon so aggressively to become number one in the pecking order, we opened the gate between the pastures and let

Marlon have the run of both. A wooden fence made with treated boards and treated four-by-four-inch posts with a metal gate separated the two pastures. The pastures hadn't changed in all these years.

Early one morning, soon after the horses had gone out to pasture, I heard a loud crashing sound. I looked out the windows and didn't see anything except two workers standing in my backyard, staring out across the pastures. Dan was in the closest pasture near the house, grazing as if nothing had happened. I couldn't see Marlon; only parts of his pastures were in view. I asked the workers if they'd seen anything.

"We heard a loud crash and saw the bay horse galloping through the pastures," they said. I walked outside to look for Marlon and saw him standing and looking around but not running. The barn worker was standing out front, wondering where the loud noise had come from. None of us could see anything out of the ordinary.

That evening, I went for my walk around the pastures. As I passed the fence dividing Marlon's pastures, I saw that a whole section of the fence was missing, and there weren't any boards on the ground. They'd just vanished. My first thought was that maybe Bob had taken them down so he could drive through with the tractor. That didn't make sense, though; he'd been driving through the open gate for years. Why would he change now? Then I noticed small slivers of wood lying in the grass. All the boards had been shattered into tiny pieces and scattered over a large area. I was stunned. Standing there trying to figure out what had happened, I knew it had to be a strong force to be so destructive.

I went to the house and told Bob. He was as concerned and puzzled about the cause of the destruction as I was. Remembering the loud crash earlier in the day, we got a flashlight and went to the barn to inspect Marlon. He was eating his hay and watched us as we entered his stall. Checking every inch of Marlon, we didn't find one scratch. We imagined a rabid animal or a snake biting him or a bee stinging him, causing him to run through the fence. We kept searching, thinking we had missed something. Marlon wondered why he was getting all this attention so late in the evening.

It was such a mystery. Something had to have scared him within an inch of his life for him to gallop from the far end of one pasture and crash through the fence as if it weren't there. He had used the gate between the pastures for fifteen years. The next day I walked the pastures, again looking for clues. A small wooded area was located at one end of his pasture, and over the years, I'd seen coyotes, foxes, armadillos, deer, peacocks, raccoons, and opossums. We often spotted a long black snake living in these pastures. Marlon had seen these animals over the years too.

We kept looking for clues but never knew what happened. I am a curious person, and I have a need to solve all mysteries. I wished Marlon could have told me what had scared him. I couldn't help but believe he could have told me if I'd been there when it happened.

CHAPTER 26
TRIP TO GERMANY AND BOB'S ILLNESS, 2009

In the fall of 2008, Bob and I went on our delayed trip to Germany. The landscape had radically changed; everything looked so perfect. These many years after World War II, there were no tell-tale signs left of the war that had raged in the area. The towns were freshly scrubbed and painted, and not a thing was out of place. They were so picturesque, they looked like they belonged in a fairy tale. A few of the towns were unrecognizable. We spent time comparing what things had looked like back then to how great they looked today. Such a different country now. The apartment buildings on top of the mountain were no longer there, but the river in the valley below and the grapevines circling around the mountain looked unchanged.

We flew into Paris for a few days before driving through France and Belgium into Germany. Having been to Paris many times over the years, we didn't notice any changes since our last visit. In Belgium, we visited the huge World

War II cemetery. It was impressive to us after Bob's twenty-nine years in the military and his time in Vietnam.

We boarded our ship in Trier for the cruise on the Mosel and Rhine Rivers. The people we encountered along the riverbanks were friendly and happy to welcome us. We relaxed, ate delicious German food, discussed the years we had lived there, and couldn't stop talking about how magnificent the German countryside looked today.

It was a great trip, but we were ready to go home and take up our horsy lifestyle where we'd left off. The Thanksgiving holiday was approaching, and we began to prepare for our family to come home. This was the first year they would be coming for Thanksgiving instead of Christmas. We were so excited to have all four college-age granddaughters coming. Schedules were different for each of them. You long for things to be the same, yet you celebrate the wonderful changes happening year by year. This year, we'd spend Christmas visiting Rob and Diane.

In January, after celebrating the holidays, we prepared for Bob's appointment at the Birmingham VA Medical Center. This was his first step toward getting hearing aids. He'd lost some of his high-frequency hearing from being in the air defense and operating ninety-millimeter gun batteries early in his military career. Over the years, his hearing had further deteriorated.

After we were married, I worked at Southern Research Institute near the medical complex in Birmingham. We lived in an efficiency apartment in a building resembling an old European rock castle in the same block as my office. I could walk out of the apartment building and cross the parking lots to the research complex. This was very important, as we didn't have a car. Bob commuted back to the university for summer school. He rode a bus into downtown Birmingham to catch another bus to the university. After summer school, he worked at Sears department store until he entered the army the following January. We couldn't wait to visit Birmingham again and see if the rock castle was still there.

Bob had an early appointment, and if all went well, the rest of the day would be free. The physical went quickly, and Bob came into the waiting room, saying he was finished. Just as we were leaving, the doctor rushed in with a slip of paper and handed it to Bob.

"I want you to get a chest X-ray before you leave," she said, "to make sure your heart is not enlarged, since you've been treated for high blood pressure all of your military career." We immediately went to a different part of the hospital, and he had the X-ray.

After a late lunch in Five Points near where we used to live, and checking out familiar places remaining after fifty-five years, we looked for the rock castle. It was still an imposing structure and looked the same. Southern Research Institute had added many large buildings to its complex. We enjoyed our day and were glad there'd been a reason for us to go back to where we'd worked and lived so many years ago.

Bob was told it could take several weeks to get the results of his physical, but early the next morning, the doctor called. "Your heart is in good shape, but there is a tumor in your right lung," she said. "It has the characteristics of being malignant; however, it appears to be operable. You should start treatment immediately."

If it had been heart or vascular problems, it would have been believable. His vascular surgery had been eighteen months before, and although he was doing great, we knew further problems could occur. After mulling over the information for hours, we began to be more positive.

"The tumor is operable and should provide years of remission time," we said. He was still active, didn't have any symptoms, and felt good. Even though we were in shock, we became hopeful.

Because of my work at Southern Research Institute doing cancer research, I realized what the long-term prognosis would be, more so than Bob did. Years ago, Bob told me while I worked at SRI and for a long time afterward, every time I stubbed my toe or sneezed, he'd panic and think, "Oh no! She has cancer." It was years before he could release that fear. That was my first career job, and I was interested and impressed at what I was seeing and learning. (Without realizing what I was doing, I came home every day and talked about what was going on at work.)

Bob was immediately scheduled to have a body scan. We eagerly waited for the results. When they came back, it wasn't the good news we'd hoped for. Lymph nodes in his chest area were involved, so surgery wasn't an option. Some preventive chemotherapy was prescribed while the radiologist prepared to start radiation. The plan was for Bob to have radiation treatments five days a week until he completed thirty-seven treatments. Weight was calculated for each person depending on bone structure, beginning weight, and other health issues. He was told if his weight fell below the minimum, the treatments would stop, and he'd be put in the hospital. He was determined not to let this happen. He wanted to finish all thirty-seven treatments, get well, and get on with his life.

Again we readjusted our thinking and told ourselves things could be worse. Bob became busy with the treatments, which helped to keep his mind occupied. Each one lasted about thirty minutes. I accompanied him to treatments to make sure all went well each day. With medical procedures and unique patients, there aren't any guarantees.

This was so difficult. I wished I knew less. Sometimes I'd look at Bob and think, "It's hard to believe, but he may not be with me much longer. How can you live with someone for almost fifty-six years, and then they are gone?" I was in crisis mode. Facing these personal setbacks, I went through every possible human emotion. I believe in celebrating the joyful and wonderful things in life as well as enduring the bad, sad, and tragic ones without complaining. Working to put this into practice was very hard to do, but as I began to dwell on all the things we had to be thankful for, it helped. We had a wonderful family, great friends, and fellow church members who loved and supported us. We'd been given fifteen retirement years, and we'd made the most of them. Sharing a full-time hobby, each having our own show horse, and competing in horse shows had brought us great joy. We'd taken trips to places we'd only dreamed about. We had few regrets, yet I wasn't ready for it to end and knew I'd never be ready. It was unimaginable.

Trying to keep us encouraged and hopeful, the doctors in the radiation department kept saying, "We're going for the cure." Hearing this was infectious, almost making us believe he could be cured. I began to feel better, although I knew what the long-term result would be.

Our job was to get Bob through these treatments in as good a shape as possible and into remission. As the treatments continued, they began to affect his esophagus and his ability to eat. Bob was determined to stay out of the hospital, and his goal was to eat as much as he could at each meal. It became an all-consuming job for me to find and prepare foods he could swallow. Everything had to be soft and bland. The radiation affected his taste. I'd find something he liked one day, and the next day he couldn't stand to smell it.

We continued to go see the horses and even ride a little, but as his treatment area became inflamed, we were afraid for him to breathe the air around the barn. Also, he became weaker as time went on. Sometimes we'd go to the barn, and he'd sit in the car, watch, and read. When Dan saw me enter the barn, he'd call for Bob. He believed any moment Bob would walk through the barn doors. I'd lead Dan out to the car so they could see each other. Dan knew this wasn't normal, but it helped. During Bob's last weeks of treatment, we couldn't go to the barn at all.

Bob made all thirty-seven treatments, and the worst medical struggle of our lives eased up. He had a one-track mind and wanted to eat good food again, and I heard pizza mentioned regularly. Finally, he was able to go out to dinner, and he had lightly sautéed scallops. They tasted so good; he believed the worst was over.

A week later, I noticed Bob struggling to eat breakfast. His right hand was hanging down, making a right angle with his arm, and he couldn't use the hand. He was so scared that he didn't mention it. After we talked about it, he explained what had happened.

"I put pressure on that hand while I was prowling through the cabinet, and I stressed it."

We were so desperate for a solution that we swallowed this fairy tale and had a good laugh. He was already going to the doctor later that morning for a sore throat. Our neighbor was taking him, as I had to stay home to wait for a repairman to come. I reminded Bob to tell the doctor about his hand.

Midmorning our friend called and said, "The doctor has admitted Bob to the hospital to treat his throat infection."

After finishing with the repairman, I went to the hospital and asked Bob about his hand. He hadn't said a word. When the doctor made rounds, I mentioned Bob's latest problem. The doctor looked concerned and immediately ordered an MRI.

"She has a big mouth!" Bob told the doctor.

The MRI revealed a brain tumor. Oh, what a setback. We felt sick. Things were going downhill fast. Everything possible was being done to keep it from happening, yet it was. Immediately, the cancer doctors came up with the next plan of treatment, and as always, they sounded hopeful that the tumor could be radiated and shrunk. They prescribed a single concentrated dose of radiation directly to the tumor. It took a week or so for the helmet to be made that would protect the remainder of Bob's brain. He was anxious to get this next treatment behind him. We still hoped for more time free from cancer problems.

The radiation was pinpointed into the medium-size tumor in his brain. Afterward he felt better, and his diet improved. We regrouped, and Bob said, "I want to go see Dan and Marlon." That day things were better.

The horses started calling to us when we pulled into the barnyard. Recognizing the sound of our truck, they gave a high, shrill call of welcome as we walked through the double barn doors. Bob visited with Dan, giving him plenty of love, while I groomed Marlon and prepared for a short ride.

"Aren't you going to get on Dan?" Cathy asked Bob.

"Not today."

"Oh yes, you are," she replied. "I am going to saddle him, and then we'll toss you up there."

Bob didn't protest, and the next thing he knew, Cathy was helping him into the saddle. Both Bob and Dan had smiles on their faces. They walked and trotted for about fifteen minutes before Bob became weary and had to get off. On the drive home, he kept talking about Dan, how happy he was to see him and to get to ride once more. Dan had also showed Bob how excited he was to share another normal riding day with him.

This was on Wednesday, July 1, 2009, and the last time Bob and Dan saw each other.

Bob's condition continued to deteriorate, and deep in our hearts, we knew what the outcome would be. On a long, rainy afternoon in the hospital, waiting for more test results, Bob wanted to talk about our years together. Normally, he wasn't one to spend time talking about the past. I welcomed hearing his version of our college days, how we'd met and married a few months later over our parents' strong objections. I especially wanted to hear what he thought about marrying a country girl who couldn't make a peanut butter and jelly sandwich, nor find anything at the grocery store to bring home and cook. Bob was worried and weak and wanted me to tell most of our story to him. He was contemplative and added tidbits here and there. Some things we hadn't thought of in years. We reminisced throughout the afternoon. We laughed and we cried, often saying to each other, "I never knew you felt that way!"

Thinking about not knowing how to make a peanut butter and jelly sandwich, I wondered out loud, "How could I have known so little about so many things?"

"You made up for your lack of ability as a cook and grocery shopper with your math-tutoring ability," he quickly reassured me. "You know your tutoring saved me, helping me to pass my math courses that first year we were married. As a result, I will forgive you for talking me into taking statistics with you." Bob always knew how to make me feel special in spite of my inadequacies.

We talked about how he'd proposed and the timing of it. With a huge grin, he began, "You didn't ask me to marry you. I have always told the children and granddaughters that you proposed to me, and I have heard them asking you if it was true. Guess I need to correct that fifty-five-year-old tale."

Two days later, coming back from having lunch, Bob collapsed as he got out of our car. It was a downhill battle for the next six weeks, including surgery. His radiated brain tumor had ruptured, paralyzing him on one side. He spent three weeks in hospital rehab. His one wish was to come home. I made all the preparations, and he was able to come home, but only for a day and a half. He never got another break from his struggles, and his earthly journey ended on August 21, 2009.

How could I begin to function now that I was alone after being half of a partnership for almost fifty-six years? I had known what was coming as time went on and he continually got worse, but where there is breath, there is some hope. Then my world shattered, and the worst happened. I was thankful that he wasn't suffering anymore. The pain had shifted to me. I was dazed and numb. Everything I did was like looking in a mirror and hardly recognizing myself—darting eyes and a sad, drawn face, trying to determine which way I should go. I struggled to keep my normal routine so I could function. Concentrating on happy times that we'd shared, visits from our children, and the many trips we'd taken our sweet granddaughters on gave me moments of relief. All the things we had been through, both good and bad, floated in and out of my mind as I tried to focus on what needed to be done.

CHAPTER 27
FIRST YEAR ALONE

I worked daily to get my grief under control so I could learn to do many of Bob's jobs. Rob helped me and gave me pointers on how to use some of my mowing equipment and explained some of Bob's other chores. "I feel like the dumbest SOB that ever lived," I said to him.

"Mom, that is called division of labor," he responded. "You'll probably discover that Dad did more than you realized."

I certainly found that to be true. Besides adjusting to facing life alone, I had to quickly grasp many things so I could take care of my five acres and our horses. In the barn, there was a farm tractor with a bushhog, a finishing mower, a posthole digger, and other attachments. My garage contained an almost-new commercial Toro Z Master mower, a large gasoline-run Weed eater, and other pieces of machinery that I couldn't name.

There were several guns in the house. Bob didn't hunt, but he loved guns and had an assortment of them. There was a .357-caliber gun in case we confronted a bear while we lived in Alaska. He had a small .22 pocket pistol so he'd be prepared if a dog attacked him while he was walking his dog Riley. There were other guns, several rifles, a shotgun, and a small CO_2 pistol. Seeing this little pistol brought back memories of the day Bob had come home with it. He was so excited.

"What is that for?" I had asked in a sweet voice.

"Remember that bird that kept pecking the side window on the truck for weeks, and we couldn't run it off? I can use this and scare it away." He had plans!

The next day, Bob set up a target using old porous wallboard and told me to come and target shoot with him. I was busy and declined; however, he was persistent. He wanted me to go first, and he showed me the bull's-eye in the center.

In a hurry, I took the gun, aimed, and shot. Bob inspected the target and said, "You missed the whole thing. Aim at the bull's-eye."

Again I lifted my arm to eye level, aimed, and shot. He took longer inspecting the target and surrounding area and couldn't find any bullet holes. "Look at the center this time," he instructed.

I quickly fired and waited. Bob rushed to inspect and noticed a chunk of the board protruding out the back of the target. He was quiet for a few moments, then said, "You hit the bull's-eye all three times!" Neither of us could believe it. We kept staring at the evidence. He half jokingly asked me if I had been practicing with the other pistols. He put up the gun, threw away the target, and our practice session ended before he shot once. I tried to hide my amusement.

Living at a dead-end road with lots of traffic to the property next to me, I decided I needed to learn to safely handle one of the guns. I'd never been to a gun shop or shooting range, but I called Larry's Gun Shop and talked to the range master. "Do you have a class that will teach me how to safely handle my gun?" I inquired.

"Oh yes, ma'am, we have two five-hour classes that will teach you all about guns—how to take them apart, put them back together, clean them."

"I only want to learn how to safely handle my gun," I said, interrupting him.

"Come to Larry's Gun Shop, get a range pass, come back to the shooting range, and I'll help you," he told me.

I dreaded going but was determined and knew I needed this information. I picked a small .22-caliber pocket pistol and several boxes of assorted shells, put them in a Books-A-Million book bag, and locked them in the trunk of my car. My plan was to stop by Larry's Gun Shop on Friday afternoon after having lunch with a friend.

About midafternoon, I rolled into Larry's Gun Shop and found a full parking lot. I finally found a parking space in the back lot as someone was leaving. I trudged up to the door, opened it, and I could not believe what I saw. Never had I seen so many men crowded into one large, noisy room. They were dressed in work or camouflage clothing, and they represented every size and age imaginable. I stood there in total shock and couldn't decide whether to leave or try to figure out what was going on. The men were looking at and handling pistols, rifles, shotguns, and ammunition and having a great time. Everyone was busy, but they stopped for a moment to stare at me, the lone lady, not gun-shop dressed and carrying a Books-A-Million book bag. There wasn't a clear-cut place for me to purchase a range pass or find out where the range was located.

Seeing a cash register in one corner of the room, I rushed over and stood nearby for what seemed like an eternity. Finally, the cashier waited on me and said the door to the range was at the far end of the room. This meant walking through the busy room from one end to the other. I stood a moment looking at the distance, and I knew running out the front door was still an option. I steadied my nerves and started making my way through the crowd. The men stopped what they were doing long enough to look and politely say, "Excuse me," then make room for me to pass. I could only imagine what they were thinking as I anxiously made my long walk to that door at the back of the room.

Inside the range area, a pleasant and knowledgeable range master greeted me. He escorted me inside a soundproof room and explained what I needed to know about my gun and the shells in my bag. "Today is opening of hunting season, and the place is overrun with hunters and wannabe hunters," he told me, explaining why it was so crowded.

When I showed him my little .22, he said, "This gun is probably too small for you." I was outfitted with safety goggles and protective earmuffs and escorted into the firing range. Before I could change my mind, I stepped up to the window and tried to fire my gun at the target. I kept trying but could not get my finger on the trigger firmly enough to pull it. I finally managed to get off a few shots, but the gun wiggled so much I never hit the target. I was frustrated and wondered if I could learn to shoot any gun.

The range master assured me, "There is a gun for you. What other guns are in your arsenal?" Telling him what little I knew about my other guns, he suggested I not get rid of any until they were appraised. He offered to come to my house and do this for free.

When family and friends heard about my trip to Larry's Gun Shop, they were surprised and wanted to hear all the details. They called me a modern-day Annie Oakley.

Several weeks later, when Rob and Karen came to visit, I arranged for the range master to come and give us his expertise on all the guns. It was an informative and enjoyable evening. He suggested my .38 revolver would be best for me.

Toward the end of the evening, the range master looked at me with a somber warning. "You must decide if you can shoot someone if it becomes necessary, and if you do, you should shoot to kill." I sat there stunned. This was heavy stuff, and I hadn't progressed beyond learning to shoot a gun. Now there was a new element added. This would take some deep thinking. On the surface, I felt I could shoot someone if my life was in danger.

Later I got my chosen gun and assorted shells and went back to the gun range. This time my dress was farm casual, and it wasn't a special hunting day. In fact, it was a very cold day, and I had the range and range master to myself. He showed me how to load my gun and explained the safety features. Then it was time for me put on the required safety gear and go into the range to try again.

I was more confident this time. I had a more suitable gun, my nerves had settled down, and I stepped up and shot. All went well, and my target sheet proved it. I left feeling good, went home, and safely stored my gun away. I hoped I'd never need it.

Months later, a loud noise jolted me awake in the middle of the night. I jerked upright in bed and had the frightening feeling that each hair on my head was standing straight up. My first thought was whether I could shoot my gun in the dark if necessary. I knew the answer, and it was no. Thankfully, my second thought was, "I recognize the noise." A few days before, I had stacked a lot of paper products in a hall closet outside my bedroom. The noise was some of them sliding off the shelf. It took a while for me to settle down and realize this wasn't the real thing. I took the advice of the range master and made more trips to the range to keep in practice.

I went to the barn as often as possible that first year. It was such good therapy for me to see Marlon. Driving to the barn along the highways, I kept my eyes peeled for the hawks sitting on the power lines or on dead tree limbs, hunting for food. Watching raptors was a longtime hobby of mine, and it was comforting to see all was normal with them, even if it wasn't with me. The drive was through rural country, and having grown up on a farm, I was appreciative of the crops planted, the rich forest, the occasional farmhouse, and the mama cows and babies out grazing in green pastures. I could smile remembering when I was a teenager and had to learn to milk our cow. I didn't mind milking, but I didn't want a boy who was interested in me to drive by and see me crossing the roads in my old clothes and carrying a milk bucket.

My outlook improved with the drive, pleasant remembrances, and anticipation of seeing and riding Marlon. I could not imagine surviving the year without my buddy. We were totally bonded with that special love and understanding a person can have with an animal. Our relationship had gone from one end of the spectrum to the other. Fear was the dominant factor in the beginning, neither of us trusting the other. Slowly, ever so slowly, the trust came. It was difficult for Marlon to trust after all he'd been through and difficult for me to attempt something new at an older age on an angry horse.

Oh, so sweet to be at this wonderful place with complete trust, love, and care for each other. Words can't express the joy we shared. It gave me hours of relief from my grief and let me begin to adjust. Marlon told me he loved my company, was so happy he could communicate with me, and felt great satisfaction in our working relationship and learning new things. Best of all, he liked getting perks at the end of the day for a job well done and knowing someone cared about him.

On a crisp fall day when the weather wasn't cold but was warning me winter was coming, I felt the urgency to make the most of it. Rain and more rain had delayed my trip to the barn for days. Both horses called me the minute they heard the car door open. Marlon was delighted to see me. I hugged him, petted him, and talked to him in my most esteemed voice. He stood perfectly still, soaking in every second of his special time. Dan was watching us and nickering to me every few minutes. He was lonely for his buddy. I went over and visited with him. Second best was better than none. Cathy was going to ride Dan while I rode Marlon. I was appreciative of her making time for me in her busy schedule.

Dan was feeling lost, as Bob had been sick all spring and summer. Knowing things weren't normal on the days Bob was able to come, Dan called him every time he left his sight. I certainly knew how much he was hurting and how lonely he was. Cathy and I tried to make it up to him.

I got my tack box, saddle, bridle, and hard hat and started prepping to ride. Marlon was relaxed and enjoying the process. He constantly turned his head, looked at me, and lightly bumped me with his nose. His eyes were soft and adoring. A horse lover couldn't wish for more. His actions said it all.

After we rode and my things were put away, I led both horses out for some extra grazing time. This was complete happiness for them. Marlon rushed me if I lingered for one second. You'd think he never got to graze by the way he acted.

Last thing before leaving, I went to the car and got the carrots. All seven horses on Marlon and Dan's side of the barn started nickering when I walked through the barn door. I treated them all, even the rabbit in its cage outside Marlon's door. Getting ready to go home, I let Marlon smell my hands and pockets over and over so he could convince himself the carrots were really gone. He accepted it well today. Walking out to my car, I said goodbye to Cathy, and then we heard the loudest kick in the barn. Without a doubt, I knew who did it.

"Why is Marlon kicking the wall saying he is angry when he was so happy?"

"He is still angry with me," she responded. "When I took him out to pasture this morning, he stopped and wanted to join Danny, the pony. I made him walk past and go into another pasture. Every time he sees me or thinks about it, he lets me know how unhappy he is."

I walked back to the barn door and looked in. He was standing along one side of his stall with his head facing the back corner. Hearing me, he looked around without moving his body, but he shifted his ears ever so slightly forward, completing his smug expression that told me how pleased he was that I understood him.

Cathy was often gone to horse shows, and there would be no one for me to ride with. After all my falls, I couldn't take a chance on riding alone. Danielle, a neighbor and friend of my granddaughters, had moved back home to continue her schooling. Before going away to college, she often rode her horse with Bob and me. I asked her, "Do you want to ride Dan and go to the barn with me?" She was thrilled to have a well-trained show horse to ride. We went once or twice a week on her off days. If Cathy was there, she helped Danielle with her riding and how to handle Dan. He was stubborn, good at sizing up his riders and taking advantage of their weaknesses. Getting special attention and having a job to do made Dan a much happier horse.

Danielle and I spent the winter months going to the barn. Then in April, Dan hurt one of his back legs. We never knew how he did it. The tendons around his hock were torn, and the leg was affected all the way into the hip area. The vet told me he had severe arthritis, and he'd never fully recover use of that leg. I was so thankful this didn't happen while Bob was living. He would have been devastated.

Dan was in great pain. Cathy and I fought a constant battle to make him comfortable. He couldn't leave his stall for months at a time. His good back leg would get tired, and he'd lie down to rest, but this often made the damaged leg worse when he struggled to stand up again. It took a long time for Cathy and me to notice any progress in his condition. We'd discuss his problems and wonder if he could ever recover enough to go home with Marlon. When he appeared at his worst, he still told us he was fighting to live and wasn't ready to give up. Even on heavy pain pills, he looked forward to feeding time and ate well. I imagined I could detect a twinkle in his eyes. Later in the summer, he improved enough to occasionally hobble outside the barn doors to graze.

CHAPTER 28
DIANE AND JULIANNE MOVE IN WITH ME, 2010

The following May after Bob's death, Diane moved back to Huntsville and lived with me while she looked for a house. She brought two cats who had belonged to her daughters before they went away to college. With them came Freddie, a large young Goldendoodle, and Abby, an older yellow Lab. Adding to the mixture, my oldest granddaughter came to live with me for the summer. She was doing a clinical rotation in Huntsville for her physical therapy degree. She arrived on Sunday afternoons and went back home to Gardendale, Alabama, on Friday evenings. Sometimes her husband, Matt, joined us for a couple of days each week. He was an engineer, and he could often work from home. It was a busy household. Diane had a new job managing an office with multiple doctors. Julianne had to work and study, and everyone plus the animals needed to be fed and cared for.

In June, Marlon came home for the first time in almost two years. He had gone to the trainer's barn when

Bob became ill. Not being able to ride alone, I had delayed bringing him back home. Dan wasn't well enough to travel. I borrowed the retired old pony that Marlon loved to play with to come home with him. The pony was a companion, but he didn't replace Dan. Marlon looked for his best friend and called him for weeks. He'd stand in the middle of the pasture, lift his head high in the air, and nicker as loud as he could.

I understood what he was going through. I was still missing my best friend, but as time went by, the fog lifted, my concentration improved, and the overwhelming grief eased some. Knowing I had to venture out more and try to begin a new life, I was somewhat receptive when my friend Madelyn told me about a mystery dinner theater show being put on to raise funds for the Burritt Museum in Huntsville, Alabama. She read me the write-up in the newsletter stating that each person attending would have a part in the play. Knowing nothing about an event like this, we couldn't imagine what the evening would entail.

We decided to go. We sent in our money, marked our calendars, and waited for our big evening out. A couple of weeks later, a letter arrived giving us a brief synopsis of the mystery play and our assigned parts. The play was set in Las Vegas, and a big-time gambler had been murdered. We would spend the evening in character, gathering clues as to who'd murdered the gambler. The shocker? I was to go as a call girl.

I called Madelyn and said, "I cannot do this!" She just laughed. She was to be a waitress. I was insistent. "Call the director of the museum and tell him I am an older lady, and I absolutely cannot play that part."

The director had no sympathy at all. He sent word for me to wear a sign saying "Retired Call Girl." That seemed worse than pretending to still be active!

As the day neared, instinct told me we needed to dress in character. Discussing a costume with Diane and Julianne, I said, "I have a dressy tea-length black skirt, black sandal heels, and a huge rhinestone cowgirl belt."

We kept trying different blouses until Diane came up with the perfect one, a rather revealing white sequined top. Considering my age, we added a gray scarf entirely made up of ruffles around my shoulders. The only thing I purchased was a pair of fishnet hose. Madelyn dressed in casual clothes and added a waitress's large white apron.

I went to the beauty salon the day of the event and announced, "I want a go-to-hell hairdo."

Bob had often said that when my hair was a bit tousled and spiky after going to the salon. The salon workers decided I also needed a similar makeup job. No amount of makeup was spared. They splashed lots of green shadow to match my hazel eyes, added enough bright-red rouge to spot me across a room, and last carefully put on black eyeliner to emphasize I was someone important.

That evening we decided I had achieved that over-the-top look. Glancing in the mirror, I knew I didn't need to wear a sign. It would be easy to guess my profession.

We nervously arrived at the museum, checked in, and got in line for drinks. Glancing all around to see if we recognized anyone, we struck up a conversation with a young couple.

"We got married yesterday!" the well-dressed young lady suddenly blurted.

We were congratulating them when she looked down the line and said, "Elvis married us."

There stood Elvis in the white rhinestone suit I'd seen him wear in a television special. Whispering to each other, we wondered if the pretend Elvis was a preacher, or maybe a justice of the peace. We kept trying to make sense of what we were seeing and hearing. Finally, we got the message that we couldn't believe a word we were told, other than it might give us a clue as to who had committed the murder.

Most of the guests were in their twenties and thirties. They had gone all out on their costumes and were playing their parts to the hilt.

The pretend Elvis told us, "I drove to Birmingham, a couple hours' drive away, to get a copy of the white rhinestone suit that Elvis wore. It didn't have enough rhinestones, so I spent hours adding more." He was animated and excited as he told us about finding his suit and the tedious job of sewing on the rhinestones.

We didn't solve the mystery of who murdered the gambler, but someone else did, and they won a prize. Dinner was excellent. The young people included us in the evening's activities, and we had fun. But driving home, we agreed one mystery dinner theater was enough for us at our age.

When I got home, the girls were still up waiting for me, wanting all the details. It was late in the evening and past their bedtimes, as they had to get up early to leave for work, but knowing it was a new experience for me, they wanted to hear how it went. The adventure let me know that I could survive an evening out and even have fun again.

Freddie and Abby liked to go to the barn with me, prowl around in the woods looking for critters, and check on Marlon and Danny. One morning, I heard a skirmish behind the barn and raced to see what was happening. I arrived in time to see Freddie snatch up an animal and toss it into the air. Afraid it was my barn kitty, I yelled at him. He threw it on the ground, and the hurt raccoon managed to run up a tree. I called the dogs, took them to the house, and fastened them in the backyard, giving the raccoon time to escape.

Going back to the barn in the evening to feed and clean stalls, I heard Freddie in the woods with another critter. Investigating, I saw that it was the critically hurt raccoon, and Freddie was mauling it. I couldn't stand this, so I rushed the dogs back to the house. I went into the woods and found the raccoon barely alive. It could not be left to suffer. I rushed back to the house to discuss the situation with Diane and Julianne.

"Get your gun and shoot the poor thing to keep it from suffering," they kept telling me.

I was reluctant but finally agreed. "If I do, it will have to be buried, and it is totally dark," I reminded them.

After much discussion back and forth, I got my gun, Diane got a flashlight, Julianne grabbed the shovel, and we paraded to the woods. They made me go first with the loaded gun. Diane was second, shining the light, and Julianne brought up the rear, carrying the shovel. It was a sad single-file march out to and into the dark woods. Diane located the raccoon from a distance and shined the light on it. I raised my arm and blindly shot. I could not tell if I was shooting high, low, left, or right. Stretching my arm out to fire, I couldn't see the end of the gun. I emptied the gun of all five bullets and didn't have a clue where they had landed. Reevaluating the situation, we knew I was too far from the target. I didn't have the heart to go reload and try again.

Leaving the girls in the woods, I went to the house and called my neighbor to the rescue. He had a large gun. He stood directly over the raccoon and shot it, quickly putting it out of its misery. Julianne had the task of digging a hole big enough to bury it. There hadn't been any rain for weeks, and the ground was very hard.

"After all this time digging, the hole isn't big enough to bury a baby chick in," she moaned.

"Wait until morning. If roaming dogs haven't carried it off, I'll take care of it," I told her.

To my surprise, the raccoon was still there the next morning. I double bagged it and put it in the bed of my truck. When I called animal control and asked where to take a dead animal, they took my address and told me to put the bag on top of my garbage can, and it would be picked up that morning.

The raccoon problem was solved, but my gun problem wasn't. When Matt heard our story, he graciously offered to go to the range with me. After another long shooting session, I was feeling confident again.

CHAPTER 29
GOODBYE, MARLON AND DAN, 2010

I did not expect the end to come so suddenly, although I knew that he was not well. Marlon, my buddy, my baby, and my friend, who'd helped me survive the hardest eighteen months of my life, would have to be euthanized. A combination of problems in both front feet became so severe, nothing else could be done. The dreaded verdict was given to me on a Monday. Marlon was on pain medicine and had foam pads taped to his feet for comfort until we could get the local veterinarian who would euthanize him together with the farmer who would bury him. His condition was too severe to even consider bringing him home. The burial would be at Cathy's farm.

When the horses came home in the spring after spending the winter at Cathy's barn, Marlon couldn't wait to get his traveling leg wraps off, pass through the gate into his pasture, and start celebrating. He'd gallop, fly through the air, and perform his acrobatic maneuvers while letting out squeals of pure joy. Then he'd switch to a fast, extended trot with head and tail held high and prance like you'd never seen before. He'd go on and on! We'd just stand, watch, laugh, and admire.

Dan was lame, so I had borrowed Danny, Ariel's old pony, to come home with Marlon and keep him company. I was very excited and couldn't wait for Marlon to get in his pasture and start celebrating. Once there, he did his usual routine, but it was very subdued. As I watched, it hit me full force that neither the horses nor me were young anymore. It was a reality check of how fast the years had marched by. All summer, Marlon was not very active. As usual, he did demand that he be first at everything: first to get his food, first out of his stall, first to be pampered. The elder pony didn't object. He was happy for the opportunity to graze as much lush grass as he wanted.

In August Marlon became lame in a front foot. I consulted with Cathy, our farrier, and the lameness veterinarian in Pell City, Alabama. The decision was made to take him back to Cathy's barn so he could be evaluated. If he needed to go back to Pell City, Cathy would take him. After much consultation with the lameness expert and trying many different remedies, Cathy was given the news that everything possible had been done. I was somewhat naive about his condition and thought each problem was being managed to some degree until the call came on Monday morning saying there wasn't anything else that could be done.

In spite of the devastating news, I was thankful for our seventeen and a half years together. Trying to cope, I kept dwelling on the beautiful relationship Marlon and I had developed in spite of such a rocky start. The first few years tested my endurance to the limit, but I had been hopeful. He'd taught me to have courage in the face of fear and how to overcome those fears. We shared great happiness and reward in our partnership. I had taught him to trust and love again after his first seven years of not being treated well. My family says I gave him horse heaven here on earth.

For years, I worried Marlon would outlive me; then who would look after an old horse? Having prayed for a solution, I was so thankful for this answered prayer, even though I wasn't ready to let him go.

Going to bid Marlon farewell, I saw that he was lined up along the wall and facing the back corner of his stall. Hearing my voice, he turned his head to look at me. Unlatching his door, I quickly walked back to hug him. He lovingly welcomed me by pressing his beautiful head with those expressive eyes up against my head and shoulder. My heart melted. Then he slowly turned on his sore, padded feet and walked to the front of the stall so he could see what was happening and get into position for his massage. Knowing what to expect when I came, he'd sigh, drop his head, half close his eyes, and let his body relax. There were many timeouts for lots of tears from Cathy and me. He had spent as much time at her barn as he had at home with me.

After finishing with the pampering and grooming, I got his bright-red championship halter with his registered name, Sonny Corleone, embroidered on the side strap and put it on him. Being dark bay, almost black in color, the red looked great on him. His eyes were not as bright as usual. He was on heavy pain medicine, but his ears were forward, and he was alert. I retrieved my camera, and Cathy took our picture. Marlon was standing with his head hanging over his half door. In the next stall, Dan was doing the same. I suddenly realized I hadn't recognized Dan, and he was calling me. Walking over to pet him, I looked into his stall and saw that he was struggling to stand on his back legs. He'd lift one foot, stand on the other foot, then quickly shift back. The injured leg was looking bad; he had hurt it again. Having been overused for these seven to eight months, his uninjured leg wasn't capable of supporting his weight any longer. The added shock of Dan's problems seemed more than I could handle.

"I haven't told you about this latest tragedy because Marlon is to be euthanized tomorrow, and I know how emotional you are. I certainly understand. I feel the same way," Cathy said.

To see Dan in such a final state of emergency was unreal and so hard to believe. Cathy and I, in conjunction with the local vet plus the lameness vet, had spent these past seven months working with Dan. Our first priority was to get him pain free. He had arthritis in the injured leg and would never be sound, but I hoped he'd recover enough to go home and keep Marlon company. The verdict wasn't so shocking, but the timing of it was too much to wrap my mind around. Cathy and I were worried Dan could get down in his stall before morning and not be able to get up. If that happened, the stall would have to be torn apart to get his body out. There wasn't a choice; Dan would have to be euthanized the next day along with his buddy of almost eighteen years. I got my camera again and took pictures of Dan.

Just before leaving, I got the carrots Marlon and Dan had been waiting for. They nickered when they saw me come through the large double barn doors into the hallway that ran the length of the barn. This alerted the horses in the stalls lining each side of the hallway. I had a large package, so Marlon and Dan could have plenty and the other horses could be treated too. When the carrots were gone, I removed Marlon's halter, signaling snack time was over. He smelled my hands and pockets over and over to make sure, then watched me unlatch the door and tell him goodbye. Once he accepted that I was leaving and there weren't any more treats, he slowly backed up a few steps, looked at me, and kicked the wall, expressing his unhappiness. Cathy and I burst out laughing

"His true personality is shining through to the end. He is still talking to me and telling me exactly what he thinks," I said.

It was an emotional ride home, filled with tears and more tears, thinking about a future without Marlon. It seemed unbearable, and all kinds of thoughts were running through my head. There wasn't any doubt about what I must do, but this didn't make it any easier. After a rough night, Diane went with me to the barn early the next day, and we visited with Marlon and Dan until the veterinarian arrived. The farmer had already dug a grave big enough for both horses. It was along the tree line behind Cathy's barn. The veterinarian arrived, again explained the process, and assured me it would be quick and painless for the horses. It wasn't painless for me!

When it was time, I led Marlon in his red halter to a grassy area near the gravesite. As we walked the short distance, I couldn't get over the feeling I was betraying him. It had taken me years to earn his complete trust. In my heart, I felt I was leading him to the gallows. This was so hard to do, even though there wasn't another option.

The vet was waiting with the fatal injection that would stop his heart. Thankfully, it was very quick for Marlon. Diane was holding Dan in front of the barn, and he stopped grazing when we left. He stood next to Diane and let her hug and pet him.

As I came back in sight, he nickered to me, asking, "Where is my buddy?" It was his turn. With Bob gone, Dan had put his trust in me.

It was so important that I be there, but my heart was breaking. Words can't begin to express the grief and deep sorrow I felt. Marlon and I had spent so many days, hours, and years together. He had learned to totally trust me, even with his life. It'll take me forever to recover from his willingness to follow me to the end of his days.

After many, many tears from all of us, Diane drove me home. It was a very emotional journey. If you are a horse lover, you know they are wonderful rays of sunshine among life's stormy paths and joyful moments among the heartaches.

Karen wrote this and posted it on Facebook: "Two trusty steeds are on their way to Heaven's pearly gates. One rider is there to greet them; one rider is left behind to write the tales of their happy trails until they meet again."

A dogwood tree was planted at the gravesite.

It was hard to reconcile myself to the fact that Marlon wasn't with me anymore. He had consoled and helped me through Bob's illness and those fifteen months after his death. I felt lost. Not a soul needed me! Then one morning, like being struck by a bolt of lightning, I had a burning desire to write my story about Marlon. It was so odd, as I'd never written anything to speak of in my whole life. I had only needed freshman English for my science major; the first-semester professor liked my essays, and the second-semester professor didn't like anything I wrote. I had to work extra hard to get a passing grade.

The University of Alabama in Huntsville has an Osher Lifelong Learning Institute for seniors, and someone had mentioned it to me in passing. The next day, after having my epiphany, I drove to the campus and searched for the OLLI office. Walking across campus, I momentarily felt like I had those many years ago; I almost skipped along, feeling young and ready to take on a new challenge.

I found the office and quizzed the secretary about writing classes. "We have a Write Your Life Story class that will start in a couple of months," she told me.

I signed up and wondered what would be required of me. I assumed it was like college English, and we would be critiqued and helped. I dreaded and looked forward to the class at the same time. I don't enjoy being corrected, but to learn and accomplish my goals, I had to work on my courage.

When the class started, I cloaked myself in all the confidence I could muster and strolled into class. Looking around, I didn't see anyone familiar.

When the facilitator started the class, I asked, "How are you critiqued?"

"It is only a story-writing class," she responded.

I wondered how this could help a nonwriter. I discovered hearing the stories read let me know what was interesting, well written, boring, or ridiculous. Some stories were brilliant, and they inspired me.

Madelyn joined me in the fall session, and John's friend Ray was also in the class. The four of us had gone to Grief Share together and had become great friends, muddling along, trying to adjust to our new status in life. We were all career military connected except Ray, and he had worked for John when he commanded Kwajalein Missile Range. Madelyn, John, and I had known each other since coming to Redstone Arsenal in late 1970.

We were having dinner one evening when the discussion turned to the writing class.

"What did you write about this week?" John asked me. Immediately he corrected himself and said, "What part of the Marlon story did you write about this week?"

I had written about how scared I was working with Marlon day by day, and how difficult it was to think about riding him. Occasionally, I was a little more optimistic until something happened to upset Marlon, and at some point it always did. Some episodes were worse than others, but all set my heart pounding as if it were trying to escape its confinement. Besides settling myself down, I had to get Marlon under control and try to convince him he wasn't in charge and that both of us would survive. The discussions continued about other horrific problems I'd endured with my "abused, mad-at-the-world, starved, type-A-personality horse" I had bought at age fifty-nine. High on the list of problems were the many falls I had endured while struggling to learn to ride Marlon.

"Hey, did you ever think about cutting your losses and getting rid of Marlon?" Ray and John chimed in. "If you bought a car and it turned out to be a lemon, you'd give it a chance, and then you'd let it go whatever the cost."

Many times this question had been presented to me, and I had thought about why I couldn't give up on him.

"In my heart, I knew he'd respond to good treatment and love," I would say. "I just never imagined it would take so long. I felt totally responsible for him after gazing into his intelligent dark eyes. He was a challenge to the nth degree, and I relish a good challenge. However, I would never knowingly have taken on such a challenge at my age. Also, I don't think about giving up until all possibilities have been exhausted. I suppose it would have taken a serious injury for me to change directions. Lastly, I'd spent all my money on this 'winning show horse,' and this was my only chance to make my dream of being someone at the horse shows come true."

My friends laughed but didn't say if they agreed with me.

CHAPTER 30

GOING FORWARD: TRIP TO CHINA, 2011

Even though my heart was broken again when I lost Marlon, he continued to provide unexpected fun moments along with the poignant ones. In September of 2011, I had the opportunity to go to China on a mission trip with Karen. She was going with a small group to work with an orphanage of disabled, abandoned children and invited me to go. Extra hands were always needed to help feed and dress the children, help them do special exercises, and especially to take them outside to the playground. Still adjusting to my life alone, I thought it seemed the perfect trip for me. I could help with the children while pursuing my dream of seeing new places and cultures.

Karen and I flew into Beijing for a few days of sightseeing, then took a river cruise before meeting our mission group. On our two-week cruise, we saw places we had only dreamed about and scenery so beautiful it was hard to believe it was real. We interacted with the friendly Chinese people and had such fun hearing them sing "Achy Breaky Heart" with a Chinese accent. We disembarked the ship in Shanghai and went directly to the airport. Flying to Shenyang, north of

Beijing, off the beaten path of tourists, Karen and I were the only non-Asians on the crowded plane.

We arrived before the other five members of our group and were met by our contact from the English teaching school for children that we would also be working with. He took us to our hotel in a low-rent, rundown part of a large city. It was located near the school and a Chinese Walmart so we could shop for the orphanage.

The contact from the American school told us we were scheduled to meet with the kindergarten parents while the children were in class. We would discuss parenting skills with them while they practiced speaking English. When we arrived, there was a large crowd of mothers present.

I said to them, "In my travels in your country, I notice that the Chinese fathers are stepping up to the plate in childcare. They appear so doting and protective of their children."

"Hey, that is all for show! At home they sit down and do nothing!" one of the ladies replied. Everyone laughed and agreed with her.

The school worked hard to be a part of the community, so our next assignment was to attend the school's Saturday evening "English Corner," held from six to eight. Many Chinese professionals in the area wanted to come and talk with English-speaking people. Also, students in the nearby college wanted to come and practice the English they had learned in the classroom.

I didn't know what to expect. A crowd was waiting when we arrived, and we were ushered into a large room with lots of tables and chairs. Sitting down at a table, I was joined by a graduate student in metallurgy, and my adventure began.

"My name is Fred, the English name I have chosen," he said, introducing himself. His English was good, but he was working hard for improvement. "I will graduate in one more semester, and I am looking for the best job available."

We discussed the similarities between my long-ago study of chemistry and his classes in chemistry today. We were joined by another young man, who worked in security. He didn't have as much confidence with his English and didn't speak up as readily as Fred, but he was very interested and attentive.

The conversation was wide ranging. Then Fred asked, "What do you do now?"

Somewhere in my rambling answer, I said, "I am taking a class in writing my life story. It's mainly about me and my horse, and how sad I still am about losing him one year ago."

Loads of questions followed. "My horse's name is Marlon, and his registered name is Sonny Corleone," I explained, and then mentioned the *Godfather* connection.

Big smiles lit up their faces as they told me they had seen that movie four or five times and loved it. We discussed my horse, why I had purchased one at an older age, and if I rode it.

"Yes, I rode Marlon for most of the seventeen years that he was with me."

Next they wanted to know the male and female names for horses. "Stallion and mare," I said. With quizzical expressions, they kept pronouncing the words over and over, finally writing them down in their notebooks.

They looked up at me and said, "Did you ride a stallion?"

"No!"

Both of them asked, "Why not?"

"They have a one-track mind," I told them, without thinking.

They kept repeating this, trying to get the meaning. Then I said it again and pointed to the word "mare." Suddenly they understood! Their expressions went from puzzlement to comprehension. After a slight blush, they started giggling. Recovering, they wanted to know if I rode a mare.

"No, I rode a gelding."

This started another long discussion. How could I give the definition of a gelding to someone struggling with the English language? Our voices became louder as we talked and laughed while trying to find the correct words for an explanation. Karen was talking to a pharmacy student at the next table, and they overheard our conversation. After much discussion in Chinese between the three guys, the pharmacy student took out his phone and looked up the definition of a gelding. Three heads were gathered around the phone to solve the mystery. With perplexed expressions, they had another conversation among themselves. As enlightenment swept across their faces, they burst out laughing. We all laughed and chatted awhile. Recovering, Fred told me, "I have just seen an American movie where the girl called the boy a wild stallion. Now I know what she meant."

I thought, "How on earth did I get into this conversation?"

When the English Corner was over and it was time to leave, both young men excitedly told me how much fun the evening had been and said they would see me the following Saturday.

They were disappointed when I said, "I am leaving to go home on Monday."

We exchanged email addresses, and when I returned home, there was an email from Fred, saying how much he'd enjoyed meeting me and that he especially liked our conversation.

Anytime China is mentioned, my experience at the English Corner comes to mind, and I shake my head and chuckle at how comical our conversation was that evening. I can still see the variety of emotions as they scudded across the young men's faces.

CHAPTER 31
JOHN, 2013

Time moved on, and John and I began going out together. In the beginning, I couldn't say two old friends were dating. After long marriages, it takes forever for it to sink in that you are now single. In addition to that, John and Konny, and Bob and I, had been close friends for the last ten years, and we had known each other for over thirty years. John lost Konny unexpectantly four months after Bob passed away. I was so stressed about losing my dear friend, I had to make a trip to the emergency room. Diane was still living in Maryland, and she called one morning after hearing that Konny had passed away. I felt dizzy while talking to her, so I lay down on the floor. The next thing I knew, my neighbor arrived to take me to the hospital. Diane had called them, as she thought I was having a stroke.

After many tests, the neurologist told me, "Your body had a glitch, but it was able to correct itself."

Adjusting to a new life in so many ways, I felt I had to become totally independent, because this was forever. Besides learning to face life alone, I had to learn many of Bob's jobs so I could continue to live on my mini farm and take care of the horses. It was a slow process for John and me, but I think he adjusted quicker.

Months into our courtship, and after a fun evening out, I got up the next morning, and it was real that our status had changed. I immediately called John and said, "I think we have gone from being couple friends to single friends."

"Yes, we have, and it is darn awkward, isn't it?" he quickly said. I certainly agreed.

John was so clever in picking wonderful and surprising outings for us. Each one was so exciting and fun for me. Sometimes I had only a hint of how to dress for an adventure that started in the morning and continued through a late-afternoon movie and dinner. One outing started with a hike on Green Mountain. I was touched, as this wasn't John's normal activity.

"You earned brownie points today, whether you intended to or not," I told him.

The small things meant so much! These wonderful moments continued.

Early in May of 2013, Madelyn and I were going on a river cruise in Russia. Before I left, John called, and in a long conversation, he told me, "You are my one and only lady. There can never be anyone else for me. I will be waiting for you. Have fun, but be safe. I miss you already!"

I was so happy, but I didn't always let him know how I felt. With my personality, I assumed if I knew, then he knew.

After I returned, the dating rumors began circulating. "Well, I guess it is official now," John quipped. The terminology seemed so different today, I wasn't sure what to call it.

"You two can call it whatever you want to, but you are dating," my daughter sternly told me.

I knew it was time to let my out-of-town family know that John and I were serious about our relationship. Talking to

Karen about John, I repeated what he had told me before I left for Russia.

"What was your response?" she immediately asked.

"I was so shocked that I only said, 'Thank you, John, for letting me know how you feel.'"

He had put a song in my heart that will never go away, yet I didn't tell him. Remembering these moments tells me why it took longer for us more reserved souls to know "when the time was right," as John often said.

Karen laughed and said, "Don't you think he would like to know how you feel?"

When Julianne was told that her grandmother was dating, she was stunned and finally blurted out, "That's weird—your grandmother dating!" Recovering, she added, "I am happy for her, but that is plain weird!"

The next time I saw her, I mentioned her comment and said, "John and I think it sounds weird too." We had a good laugh, but I kept thinking, "If that was my grandmother telling me she was dating, there aren't words to explain what I would have thought. Shock and disbelief would be two of them."

My biggest surprise for John was the birthday party I gave for him in July that summer. When we were out to dinner one evening, the conversation turned to where to get great crab cakes.

"Diane makes the best ones," I told him.

"Maybe she will make them for me sometime," John said.

For the party, Diane made crab cakes. Madelyn helped out by adding a special dish to the meal. It was a wonderful evening, and John loved it, even though he always said, "I don't want a birthday to-do!"

A few nights later, we were having a romantic dinner at our favorite restaurant.

"John, I have a question to ask you, and I need a yes or no answer," I said as I reached across the table and took his hands in mine.

The look on his face was priceless. "Can I give you an answer by morning?" he responded after a long pause.

"No, I need it now!" Without further ado, I said, "I have always heard that the way to a man's heart is through his stomach. I pulled out all my tricks for your birthday dinner, even getting Diane and Madelyn to help me. I want to know if I accomplished my mission."

There was relief in his expression as he laughed and said, "You sure did a good job of setting that up!" I loved teasing John. I never remembered to ask him what he thought my question would be.

In August, John went to visit his daughter Ashley and family in Auburn, Alabama. He called early in the morning before he left and found out I felt sick and was going to the doctor that day. He called daily to see how I was doing and to pass on "our" doctor's advice for similar ailments. John had convinced his doctor to take me as a Medicare patient, even though he wasn't taking any more.

Returning over the weekend, John called and said, "When you're feeling better, I want to come to see you, and I am bringing dinner."

It was Tuesday before I was feeling decent. John came with a mission. He brought dinner, a few *Wall Street Journal* articles, and some magazines for me to read.

I took some of the things from him and was putting them away when he suddenly said, "We need to talk!"

He had said this before and often ended the conversation with, "when the time is right."

"OK," I said as I continued laying things on the table.

"We have to talk now," John said as he walked closer and turned me to face him. Before I could respond, he said, "Will you marry me? I love you! I have dragged my feet long enough, and we don't know how much longer we have to be together. I love you very much, and I want you to be my wife. We can blend our lives, our families, and our houses. We can live in two houses until we have time to settle things." He paused. "Well, is it a yes or a no?"

I was so surprised at the timing of his proposal that I hadn't said a word. "Yes! Yes!"

"I want to marry you before your birthday in May," John added with a mischievous grin.

A short time later, the wedding date was moved to the first of the year. "It is like the dam has burst," he said. "Everything I ever wanted to say to you just comes pouring out, so easy and natural."

I felt the same. As the hours and days went by, our hearts just exploded and overflowed with love for each other. We talked all evening and laughed over some of our funny courtship clues that we had sent each other. It was a wonderful and magical evening, and we were never the same since. Guess we were surprised that at our age we could be so totally in love, even feeling giddy and silly, and we couldn't wait to spend the rest of our lives together.

We went to dinner a couple days later to our favorite place, and we wanted to tell everyone, stranger or not, "Hey, look how happy we are! We're in love, and we are getting married." John had planned on proposing that evening, but once he saw me on Tuesday, he said he couldn't wait any longer.

We couldn't tell anyone our news until his daughters, Caroline and Ashley, were told. Caroline was preparing for minor surgery and had an important work assignment. John didn't want to distract her with wedding details until these were finished.

The following Monday were the Labor Day festivities at Redstone Village, where John lived. He invited our friends Ray and Madelyn to join us for the celebration. It was a wonderful and busy day. John was leaving on his long-planned trip to New York the next morning.

That evening when he called, he said, "I am delaying my trip for one more day. I need to see you. Yesterday was not a proper goodbye. I am coming tomorrow evening and bringing barbecue for dinner."

Besides the barbecue, he brought a list of things that he didn't want to forget to tell me, and he wanted to emphasize over and over that his house would totally become our house.

"We will do whatever it takes to make it so," He said.

John looked up from his list and said with intensity and passion, "You are on my mind every moment of every day and every night."

To me this said everything! He didn't have to say another word. He had expressed so elegantly our deep feelings for each other.

We talked and planned mornings and evenings while John was away. "I am not traveling again without you," he said. "We need to get married by the first of November. Each year, I attend a conference with my daughter Caroline in late November, and I want you with me."

The Big Bay Horse Doesn't Live Here Anymore

My family was coming to the farm for a week the first part of October. I knew they couldn't return in such a short period of time. John and I talked it over and decided to get married at noon on Saturday, October 12. This date depended on his daughters being available to come.

The last evening John was in New York, we talked long into the night. He kept calling back as he thought of more things he wanted to say. We firmed up our wedding plans.

"I am ready to come home, but I am obligated to make a brief stop by the military reunion in Maryland," he told me.

On his way the following morning, John pulled to the side of the freeway and called. "I have an easy six-hour drive, and I will call this evening. Just want you to know how much I love you!"

I got home late from shopping for a wedding outfit, and when the phone rang, I thought it was John. Instead, it was the beginning of my nightmare.

"Dad stopped at a gas station along the freeway and he fell—"

"Is this Ashley?" I asked. She was crying, and I wasn't sure who was on the phone.

"Yes, this is Ashley! Is this Nell?"

Both of us were so upset, we questioned each other to make sure we understood who we were talking to. "Is he in the hospital? Where? Is it his heart?"

"No, he passed away," she said.

The disbelief, shock, and feeling of not knowing how we could go on without him took over our minds and hearts as we tried to process this.

"It can't be real. I talked to him a few hours ago, and he said he'd be there by midafternoon," I quickly said.

"He stopped at a convenience store and was walking back to his car when he collapsed and died instantly," she told me.

"I am so thankful he wasn't driving," I managed to say to Ashley.

"Oh, my goodness, I hadn't thought of that!"

We finally hung up in total disbelief. I was so distressed that Diane came and spent several days and nights with me.

My family had been informed earlier that morning that John and I had set the date for our wedding, and later in the evening, they were notified that John had died. Rebecca, my youngest granddaughter, was so upset that she couldn't sleep. Late in the night on her cell phone, she began writing down all her thoughts and feelings she was having about what had happened to Mama. The next morning, she sent this poem to all the family.

"I can't believe this! When did she have time to compose it?" I kept saying over and over.

"THE GUARDIAN ANGEL, 9-11-2013"

My Mama is one of God's guardian angels.
He placed her on this earth to bring happiness to others,
To provide strangers and loved ones with caring and understanding,
And to help the weak lift their heavy loads.
Her life, her story, is a testament to her enduring strength and endless love.

The Big Bay Horse Doesn't Live Here Anymore

She was married to her husband for over
fifty years.
She was a good wife to him, loyal and patient.
They raised two children together
And traveled the world together.
Her husband was strong in many ways,
But his strength was no match for my
Mama's.
He would have been lost without her:
He would have been a ship with no light to
guide him.
So God took him first, because He knew my
Mama could shoulder the pain.
She was his guardian angel and let him pass
her on the way to heaven.

She was left with two horses that she had
cared for and loved for twenty years.
She had struggled with them, cried with
them, and triumphed with them.
They brought her comfort and reminded her
of an easier life,
A life before she was alone in a big house on
a big farm.
Her horses were growing older, and so was
she.
She knew that they would not understand if
one day she were gone.
They would look for her and wait for her to
come until the end of their days.
She made the difficult choice when it was

their turn to go:
She shouldered the burden and stayed with
them so they wouldn't have
to be alone.
She was their guardian angel and let them
pass her on the way to heaven.

My Mama relied on her friends, her family,
and her faith to pull her through.
She was strong, so she carried on.
She developed a deep friendship with a man
who had endured similar tragedy:
his wife had passed shortly after Mama's
husband had.
They enjoyed each other's company and
relied on each other
when life was too much to handle alone.
Her heart opened once more, and she realized that she had fallen in love with this
man.
This man loved her too,
So much that he could not even bear to
leave her behind when he traveled.
They both knew they were getting older and
would not have a great many years together.
God knew this man would be crushed if
Mama were to leave him behind on earth,
So God took him first.
This man was able to live his final hours in
hope and happiness
while my Mama shouldered the pain.

She was his guardian angel and let him pass
her on the way to heaven.

My Mama deserves the best that this world
has to offer,
But she is handed heavy burdens instead.
All because she is a guardian angel
And would rather suffer all by herself
Than force someone she loves to take up any
of that heavy burden.

When her time eventually comes
To walk into the kingdom of heaven,
God and His Son,
And all of His angels,
And all of the thousands of people
Who had their burdens lifted by my Mama
Will welcome her and celebrate her.
She will finally have the pain and suffering
lifted from her shoulders,
And she will live her eternal life in absolute
joy,
Surrounded by pure love.

"This came to me as fast as I could write it down." Rebecca explained later.

I relived that evening of September 10 when the nightmare began, over and over. As time eased the pain, I could concentrate on all the wonderful things about John: our long friendship, our deep love for each other, and the miracle that we believed had happened to us late in life. I was comforted that he went as he hoped he would. Also that we'd had the opportunity to express to each other exactly how we felt and to share in the joy, excitement, and happiness in making future plans together, even though our time was far too short.

We weren't married, but we had the bond that married couples have, sharing all our joys, sorrows, and life's daily happenings. We had so much in common and complemented each other. I am the curious one, and I always want to know the answers. John felt there were times when it was best not to know. I like challenges and adventures; he did too, but he believed that safety considerations came first. I believe in getting the job done. "You just push right on through," he told me again and again. John took a more mellow approach to the smaller problems in life.

I ran across this quote by Elizabeth Kübler-Ross and David Kessler: "The reality is that you will grieve forever. You will not 'get over' the loss of a loved one; you will learn to live with it. You will heal, and you will rebuild yourself around the loss you have suffered. You will be whole again, but you will never be the same. Nor should you be the same, nor would you want to."

Seeing this quote after John's death, I understood it. These many months have been long, painful, devastating, and unbelievable. The shock of his death, and the deep sorrow of losing him so soon after we had reached the mountaintop in our relationship, were too much to bear. It was so difficult for me to accept. I listened for the phone to ring from morning until night. My heart jumped when it did ring, only to be disappointed again and again. I wanted to talk to him, to tell him my latest news, to ask for advice, to discuss our common interests, to go over the day's activities—just to hear his voice

A year passed before I could accept that he wasn't ever going to call again. The grief was all consuming, and I couldn't escape it. I had to find a way to deal with it. I was never bitter or thought I was immune to tragedy. It just hurt so much, and there wasn't any relief. He was still on my mind when I went to sleep at night, and he was the first conscious thought to flit across my mind when I woke up in the morning.

As the shock lessened, I wanted to hear love songs. They couldn't be any old love song; they had to say exactly what I wanted to hear and to completely fit my situation. I could listen to the Jersey Boys sing "My Eyes Adored You" over and over. Then I'd hear another song that expressed my feelings for that day, and I'd start listening to it. On and on it went.

One day, I got in my car, turned on the radio, and heard George Jones singing "I'll Never Get Over You Until the Grass Grows over Me." I was speechless! He was singing my song. I adopted it. I began to listen to other George Jones songs. Oh, how he felt my pain and expressed it so perfectly with an authentic voice. At last, someone was putting into words what I could not. On better days, I could go back to listening to love songs that were more uplifting.

A couple of years later, I was driving through my neighborhood with all the dogwood trees in full bloom, and I felt real joy bubble up for no reason at all. I almost stopped the car and celebrated. As the anguish lessened with time, I became more at peace about losing John. I still think about him every day, and about all the wonderful time that we shared.

Yes, I am forever changed. Not a day goes by that I don't think of my departed loved ones. I am more appreciative than ever for my loved ones who still share my life.

I am thankful to all the people who encouraged and helped me along the way!

Bob for having the dream, and letting me share it!

Bobby and Greta, for helping me remember our early farm life!

Writing Your Life Story friends, Betty, Madelyn and Ray for being there in the beginning.

John Jeter's Writing Group, for pushing me forward.

Rocket City Writer's Group, the name later changed to Huntsville Writer's Group, for critiques given with kindness and love.

Coffee Tree Writers Critique Group, for valuable suggestions and corrections.

My readers, editors and special helpers: Jennifer, Stacy, Kim, Nina, Rhonda, Ann Marie, and Bobby, for your help and believing that I could do this.

My love and admiration to Cathy Herrin, my friend and the best Horse Trainer and teacher in the world. I am forever indebted to you. I appreciate your patience, tough-love, and your sense of humor in teaching this older lady to ride. It was a terrifying, long and an almost impossible challenge, but your encouragement kept me struggling onward.

To my wonderful family: Rob, Karen, Diane, Bruce, and my four granddaughters, Julianne, Jessica, Jeannene and Rebecca, for sharing all mine and Marlon's years of disasters and triumphs. I appreciate your love and support, even while worrying about my many falls. Especially, for listening to me talk about Marlon's unbelievable shenanigans. I love and appreciate you!

ABOUT THE AUTHOR

Nell Parsons grew up in a farming community on Sand Mountain in North Alabama. She attended the University of Alabama in Tuscaloosa and majored in Chemistry where she met Bob in a chemistry lab. Later, they married and spent the next 29 years traveling through the United States and other countries with their two children. They retired in Huntsville, Alabama. For their retirement hobby, they each purchased a show horse. It turned out to be a long, terrifying, but rewarding adventure for Nell. Getting up one day, she felt an over whelming desire to tell her story.

Made in the USA
Columbia, SC
08 August 2020